THE Nesting Season

THE BELKNAP PRESS *of* HARVARD UNIVERSITY PRESS

# BERND HEINRICH

## THE Nesting Season

CUCKOOS,

CUCKOLDS,

AND THE

INVENTION OF

MONOGAMY

Cambridge, Massachusetts | London, England

First Belknap Press of Harvard University Press paperback edition, 2011

*Library of Congress Cataloging-in-Publication Data*

Heinrich, Bernd

   The nesting season : cuckoos, cuckolds, and the invention of monogamy / Bernd Heinrich.

     p.   cm.

   Includes bibliographical references and index.

   ISBN 978-0-674-04877-5 (cloth: alk. paper)

   ISBN 978-0-674-06193-4 (pbk.)

   1. Birds—Behavior.   2. Courtship in animals.   3. Parental behavior in animals.   I. Title.

   QL698.3.H435   2010

   598.156′2—dc22    2009049533

# CONTENTS

# THE Nesting Season

# INTRODUCTION

*All nature is but art, unknown to thee;*

*All chance, direction, which thou canst not see;*

*All discord, harmony not understood;*

*All partial evil, universal good:*

*And, spite of pride, in erring reason's spite,*

*One truth is clear, WHATEVER IS, IS RIGHT.*

Alexander Pope, *An Essay on Man*

I WAS KEEPING AN EYE ON A GOSLING AT the edge of our unkempt lawn near a large beaver bog today by our home in Vermont. Normally the real parents have five or six progeny. This nearly two-week-old gosling has known me and the other family members since it hatched fully formed and ready to roam the second day out of its egg, on May 10. I was regretting having taken on the very demanding job of parenting a bird that requires very little from its parents, relative to most species. Sadly, I had already failed, because yesterday I lost the second of the two goslings when I went into the house for only five minutes. I had left the two to their own devices, and they then peeped in frustration at my absence. In minutes a raven that has a nest with five hungry young on a nearby cliff took the opportunity to swoop in, kill, and fly off with one of my charges. This traumatic event was a vivid reminder that goose parents or alloparents can never, not for an instant, allow themselves the luxury of taking their eyes off their young, nor can they permit the young to stray away. Both of my own kids were attached to the goslings, and they cried when they found out that "Lem" was killed during my turn as alloparent. I was distraught as well. It turns out that we aren't adequate parents for Canada goose goslings. It's a much too demanding job for us, though these birds can feed themselves from the day after they hatch. The main requirement of the job is a bonding that is so strong that no separation is tolerated, which pretty much applies to our parenting as well.

The remaining gosling, named "Clover" by the kids, insisted on being with me constantly; I could not leave it even for five seconds. If separated from me, it made pitiable-sounding drawn-out, upward-inflected peeps that would melt your heart. I had to get away, because otherwise this baby would drive me crazy. On the other hand, once we were reunited, it made sweet and gentle chatters and purrs so soothing that I could listen to them all day long. Of course these vocalizations are normally meant to be understood by the gosling's natural parents. They aren't intended for communication with a primate, but that's what they did.

The first peeps mean something like "I'm alone, come find me," and after you return they mean "All is well with me." The wonder is not that they can communicate without words. It is that seemingly arbitrary sounds are not arbitrary at all—they touch us emotionally at some deeper level than rational understanding, and the understanding can transcend species. There were lessons to be learned from the goose fostering, and for me the most significant ones were the obvious, that a strong mutual attachment had been formed between geese and humans, and that the bonding was mediated by a "language" that was either very quickly learned or so quickly learned that it has instinctive adaptive underpinnings.[1]

The goslings have evolved to maintain constant contact with their parents or alloparents, and their parents have evolved to maintain the same contact with them. The young induce the parents to act appropriately by crying. After the stress of separation is over, the young switch to a different vocalization, one that signifies their satisfaction. And how well I remember the same thing with my own babies—the long "waaa's" and the sweet cackles! I understand the baby geese's calls because the analogous calls induce in me the same behaviors they induce in the goose parents. And it is not only these vocalizations that touch and move me. I also understand their language of gestures enough to understand when the young gosling is relaxed and when it is frightened and tense. But while this may be interesting, what makes it significant is that when I contemplate the sounds and the gestures of insects, crustaceans, and arachnids, I may, after rational analysis and experimentation, perhaps be able to figure out what they mean, but I cannot empathize emotionally. I have no way of knowing what if anything these organisms may feel. I only see responses to stimuli that solve their problems of living, and I feel as though I am confronting a fascinating but alien life form that faces many of the same fundamental problems that all life must solve to survive and reproduce. Yet it seems so alien that no emotional bonds are possible.

These "others," which to a lesser extent also include numerous vertebrate animals, provide contrasts that highlight our commonalities with birds. All other mammals are closer relatives to us than to any bird, but birds and humans share many similar parenting problems, and so we have evolved some similar or evolutionarily conver-gent solutions. Perhaps we are therefore more like some birds despite having more distant evolutionary connections to them than, for ex-ample, cows or moles. Those evolutionary connections based on de-scent, which are the basis of taxonomic classification, seem to almost define the only kinships that we acknowledge. In this book I hope to show more.

As a physiologist by training, my outlook is informed by com-parison of mechanisms. Despite specific adaptations designed for spe-cific circumstances, the same scaffold of mechanisms of digestion, circulation, respiration, endocrines, and neural function operate in birds and in us. When the behaviors are so closely identical that they can easily be understood by us, and sometimes vice versa, then the physiological underpinnings are also likely comparable; that is, the basic similarities are probably more than a mere presumption. For ex-ample, there must be a mechanism that motivates young goslings to be attracted to their parents, and vice versa. Without these mutual at-tractions in this species, there would be a failure in reproduction and the young's survival would be seriously compromised. The attach-ment is real, and until shown otherwise I do not feel apologetic for calling it "love," in part because that is the term for these behavioral adaptations that is best understood by us. And as I will briefly explore later, there is physiological evidence that in birds as in mammals it in-volves the identical neuron-endocrine axis.

The lives of birds can help us to better understand ourselves. I am at times willfully anthropomorphic to draw attention to such sim-ilarities, although I also highlight birds' often seemingly bizarre dif-ferences from us and from each other. Recognizing that the diversity among birds and between humans and birds stems from common problems of parenting under the demands and constraints of diverse environments may be meaningful in understanding some of our own diversity.

Birds are extraordinary creatures by almost any measure. But they are especially impressive when we are so brash as to compare them to ourselves in their astonishingly diverse ways of becoming parents and of parenting. We need three or four decades to accomplish what birds of many species routinely accomplish in less than a month—court, mate, build a nest, lay from one to about twenty eggs, incubate them, and then feed and protect the young to adulthood. In some cases birds also provide for their offspring's education. They may have to migrate tens of thousands of miles just to start the nesting season. Birds are arguably born three times: an egg is laid, the egg is hatched, and the young bird fledges. Not surprisingly, the whole elaborate process of parenting is so time-consuming and requires the procurement of such enormous resources that for most birds it demands a partnership of both parents, and for some even that is not enough.

In all of their activities birds are, in their infectious exuberance and beauty, a vital link in Earth's ecosystem. Many exist only in the remotest, most exotic places while some now live almost exclusively in association with humans. They are all around us, offering a psychic link to nature both through their beauty and the fascinating complexity of their lives. Furthermore, as they enhance our lives through their beauty and their majesty, they have provided us with many of the keys that help us to understand life.

Birds have entranced me since at least the age of six, when for four years I had a pet crow, a jay, and a wild pigeon, all of whom came from nests I had found. Between the ages of 8 and 10, shortly after WWII, I lived with my family in a hut in a forest in northern Germany. My days were filled with the excitement of hunting birds with a slingshot and finding nests. The birds were sold to museums, but I kept an egg from each species for a collection that opened my eyes to wonder then and provided the subject for a watercolor painting now. Since those days over sixty years ago I have lived with crows, a great-horned owl, ravens, and the geese with whom I still share affections. I have never lived anywhere without doing two things: planting several trees and building and putting up some bird houses. In the early 1960s I took a little over a year away from my college studies to travel into the wild and pristine areas of East Africa with my parents, who collected birds professionally. It was only much later, however, that I learned about the insights into biological processes that had come from the scientific study of birds, and these then opened my eyes from the exterior and sensual beauty of birds to a new and unsuspected hidden world of evolutionary processes. My hobby was drawing and photographing birds, and without them I would have been interested neither in art nor in science. In a way, I combine it all here and reach back from my emotional roots to the scientific "tree" of knowledge that grew from them.

The scientific literature on birds and their reproductive and nesting behaviors is enormous and potentially confusing. For over a century, year in, year out, new findings have been published. The volume of the technical scientific literature is intimidating—it fills libraries, and no one person could absorb it all. I feel very apologetic for not having more than a passing knowledge of some of the exciting research that has occupied some of my colleagues for a lifetime, yet I want to at least give a flavor of selected aspects, even if it means that I may be at times more technical than some readers may care for and not enough for others. Although we know an increasing number of details, perhaps we lose sight of the forest, and there are still many unexplored thickets. For example, most birds camouflage their eggs by coloring them with spots, blotches, and squiggles, and hole nesters (for the most part) have white, uncolored eggs because they don't need to hide them. The young that are fed in the nest gape and beg in response to mechanical stimuli, to be fed (usually) by their own parents, and the parents of most of them are meticulous about eating or carrying away their chicks' wastes so that the nest remains unsoiled and does not attract predators. However, even as we unravel the function of hundreds of common behaviors and publish them in thousands of publications, there are always anomalies and questions. The mourning dove pair with their flimsy platform of a few twigs that I watched through a breeding cycle on the apple tree next to my house has pearly white eggs, the young don't gape, and all of their feces accumulate in the nest.

Despite all of the science entombed in research volumes that fill libraries, birds remain amazing and are often mysterious. Why are the eggs of the marsh wren deep dark brown, a winter wren's almost white, those of the gray catbird brilliant greenish-blue, those of a phoebe pearly white, and those of a red-winged blackbird spotted, blotched, and squiggled with brown, lilac, black, or purple? We don't really know how an albatross can be displaced a thousand miles of ocean away from its nest, yet still return to it directly. We still don't know precisely how the instructions are inscribed on the DNA in a cell of a penduline tit so that the male weaves a domed nest of fibers and the female lines it with a thousand feathers, whereas a bird-of-paradise male builds no nest at all, and a bowerbird male builds an elaborate structure like a house that he decorates to attract females to mate with him. I hope to shed light on some of these questions, not only by a perusal of the scientific literature but also by my original observations, thoughts, and speculations, illustrated with my photographs and watercolor sketches.

Almost everything about birds that excites relates, directly or indirectly, to their reproductive behavior. Here in the north temperate world, spring is almost defined by the birds' return and their nesting season. And nesting encompasses just about everything about birds that grabs our attention—their varied and colorful plumages, their songs and spectacular displays, the varied sex roles and cooperation among mates, monogamy, extra-pair copulations, polygyny and polyandry, various parenting roles, helpers at the nest, nest parasitism, predation and predator avoidance, infanticide, egg coloration, cognition, and deception of predators. I believe that the more we understand the diversity of birds and the problems and solutions of their lives, the more we can understand and perceive the wonder and the beauty of the whole Creation. Socrates famously said that an unexamined life is not worth living. If so, then perhaps the examined lives of birds make it worthwhile for us to preserve *their* living, so that we will grant them the vital habitat and toxin-free environment they need in order to thrive.

I

## LOVE BIRDS

MARCH 14, 2006. NEAR 3:00 PM I HEARD
a wild, excited screaming of a broad-winged hawk, still at least a half-mile away but approaching rapidly and increasing the vigor and apparent excitement of its calls. When I finally saw it, it was alternately folding its wings to the sides, diving, and then pulling up and calling again. It kept repeating the diving and calling as it headed directly to its former nest site, where later the pair

nested again. I may have witnessed the final leg of the return journey of one member of our resident pair.

Every year for at least the last five years the pair has built a new nest about a hundred yards beyond our house, thirty to forty feet up in a triple fork of a broadleaf tree—usually a maple or a birch. Though the hawks are here all summer, I never see such a show of exuberance. Presumably the bird today was joyfully excited, as one might expect from the mechanism that draws them back to this one place in a vastness of thousands of square miles of forest. Although both these woods and the bird's potential mates are not special as such, they are special to each other.

They say that love is blind. With male bowerbirds and birds of paradise that is probably true because they mate with any female that chooses them, but it is certainly not true for the females. They choose only the most splendidly plumed male with the best sound and dance repertoires or the most elaborate bower that they find. Proof of the females' discrimination, not just now but over millions of years, lies in the spectacular creatures that the males have become through sexual selection. In monogamous (if one refers to the usual male perspective) species other criteria besides displaying beautiful plumage along with song come into play in the choice of partner. And those criteria are often not easy for us to fathom. One perhaps extreme example of uncritical mate choice is a report in *Audubon* magazine (March-April 2008) of Petra, a black swan from Australia who bonded to a white plastic swan-shaped two-seater pedal boat. Petra is the only black swan on the Aasee Lake in Munster, Germany, and she "stares longingly at the vessel and swims circles around it." She bonded with the one she was with. Black swans don't migrate, and when winter came, the swan and her pedal-boat "boyfriend" were transferred to the nearby Allwetter Zoo, where I assume they will live happily ever after.

To suppose that a bird "stares longingly" or even lives "happily" is to open oneself to the criticism of committing an anthropo-morphism (I'll abbreviate to APM), which is defined as attributing unique human emotions and sensibilities, motivation and reason, to a nonhuman animal. Applied uncritically, it is a deadly scientific sin, one that often displays ignorance. For example, on seeing a moth tremble when touched, we might say it is shaking from passion and excitement when in fact it is shivering to warm up its flight muscles in readiness for flight. In the case of Petra, however, the attraction is de-monstrable, not a figment of the imagination. Precisely what Petra is feeling may be unknowable. But "love" is probably not far off the mark since the word conveys a common body of facts of physiology and behavior that encompasses specific behavioral and physiological criteria, so why not be objective and use it for others beside our-selves?[1]

Anthropomorphisms are eloquently discussed by John Alcock in his book *The Triumph of Sociobiology*.[2] My purpose here is not just to paraphrase previous arguments, but to give you my take on the topic, because I do not necessarily shy away from some inferences that might be construed as anthropomorphic. I want to tread the fine line between differences and commonalities between species, whether between us and birds or between different species of birds. In practice that goal results in having to differentiate between blatant, mild, and apparent APMs with the distinction between them being determined by current knowledge, which is always incomplete. I shall address the last and simplest first.

The "apparent" APM is shorthand jargon used by evolutionary biologists and is used by them to save long explanations. It may sound like an inference based on attribution of human volition when we de-clare that a "bird incubates its eggs to hatch out its babies." The oper-ative word here is "to." It sounds as though we meant that the bird is consciously trying to produce babies and knows it has to keep them warm so that they would grow properly, and so acted accordingly.

But it is extremely unlikely that a bird has any notion whatsoever what it is doing by all of its elaborate behavior of building its nest, laying eggs, and incubating them, any more than it could know (or care) why it mates. Saying "to" is just simpler than explaining that birds incubate because in their evolutionary history there have been many stages of birds' closely associating with their eggs, and those individuals that associated with them the most closely had slightly higher reproductive rates than those who didn't, and that behavior therefore eventually became inscribed on the genes in the population, and it came to be activated under the influence of hormones in response to the stimuli, and so on. Such explanations become extremely tedious if used simply to avoid the word "to." They would need to be repeated almost exactly with every single apparently purposive behavior that is mentioned, one after the other, to complete the story.

The "blatant" APMs are those inferences that ultimately derive from *not* knowing anything about the animal in question and taking a shortcut to try to fit an "explanation" to account for its behavior strictly from our own sensory, rational, and emotional worlds. The tricky part is that all of these worlds—theirs and ours—do overlap to a greater or lesser extent between many animals. Indeed, the stigma of APM is greater depending generally on the distance of genetic relatedness and life strategy that is separates them from us. Since modern biology is revealing increasingly more commonality between us and many other animals and emphasizing similarities, there is an increasing lack of clarity between valid inferences and APMs, and therefore increasing risk of ignoring the obvious.

To illustrate this, consider the following examples in birds. It is an APM to suppose a bird sings because it is happy. Most probably it is; it is unlikely to sing if it is not vigorous and healthy, so the song advertises its health. It sings also under the proximate influence of testosterone, which has the effect of elevating mood in ourselves as well. Birds sing ultimately (the evolved or functional significance) "to" advertise themselves, hold a territory, and get a mate, and once they have both, then singing greatly diminishes. So are they then less happy? Maybe they are, but while happiness may permit, it does not dictate. The "because" also encapsulates a train of logic that appears to be anthropomorphic and would be cumbersome to reiterate, but it is understood that there are proximate and ultimate reasons, and volition belongs to the first and is totally a separate issue from the evolutionary reason why birds (have evolved to) sing. It may be correct that the happy bird sings because we can reasonably suppose that an unhappy bird has something wrong with it, or it is missing something that prevents it from singing in otherwise suitable circumstances. But whether or not it is "happy" is not as much a suitable scientific question as why it has evolved to sing and why it sings under specific circumstances.

Some APMs may be more justified than others. We would feel anguished and sad if one of our children were suddenly missing. We would hunt frantically; there would be visible behavioral changes that we would recognize if we saw them. When we see the same behaviors in another species, we infer the same or a similar basis for them. But we can't safely extrapolate our perceptions without evidence, such as if, for example, one were to take one of their brood from them. Most birds indeed act disturbed if their nest is approached. However, if we were to remove an egg or a young bird surreptitiously and undetected, then the bird is unlikely to change its behavior, so if that is indeed the case, then it is reasonable to suppose the removal was not missed. If in doubt, then one could do more tests. At the same time, it is standard operating procedure to invoke the scientific approach (based on countless research results) and safely suppose birds "try to" increase their reproductive output and minimize loss "in order to" have as many young as is safely possible. But this is for the ultimate, or evolutionary, "reason" and has no bearing on their perceptions. We as well as birds may need "love" to cement a long-term relationship, but like animals other than ourselves we don't need knowledge of consequences in order to perform the adaptive agenda.

*June 12, 2009.* It rained during the night, and this morning it is still sprinkling while I'm at Sunset Lane Pond to watch the geese. As in many previous years they had again gathered here from miles around. The open fields permit good grazing, easy views, and quick access to the shore of a big pond. This spring the total aggregation consists of five pairs of adults and a total of twenty-seven young, who wander around grazing together as a mega-group composed of four separate "families." Although in some years one pair has led the young of several parents, this year there is instead one group with two pairs of parents and four family groups in all. This morning I arrive as they are starting the day.

Some of the adults are standing and look alert while all the young with them are lying down in piles with their heads into their shoulders, still sleeping. They are at first packed too close together to count, and I am not sure where one group ends and another begins. It is an idyllic-looking pastoral scene that I watch from my pickup truck parked along the road above, less than 200 meters distant.

I had picked the right time to watch, because in ten minutes, first one and then another gander starts doing head shakes to get his group up and going; the head shake means "let's go." The geese of the big two-family group with twelve young (half are still all yellow fuzz and the other six already show a little black and white on their faces) start to walk slowly, and suddenly several of the gangly older youngsters jump up and run ahead, making their peeping sounds and flapping their wing stubs, which are starting to show only the first signs of pinfeathers. After their brief sprint they stop and start grazing on the short, already cropped grass and clover. The rest of the group catches up. This group of twelve young and four adults then works its way over to join the next couple with six young. The gander of the third group shakes his head, and then that couple with their six also joins the now rapidly growing crowd. The fourth family with three young is slightly off to the side, and finally rouses itself to join up with the rest. The whole crowd of thirty-seven geese—ten adults and twenty-seven young—now slowly advances like a giant amoeba. The geese work their way around the pond, feeding obliquely to me, the young almost all with their bills down and grazing and the adults with heads high and alert. Once in awhile one of the adults rises on its feet and flaps its wings, and I notice blue stubby pinfeather shafts; these birds are no longer flight-capable because they are missing their flight feathers and just starting to grow new ones. Later they all enter the pond, and now the four groups—still all moving together—most clearly sort themselves out into families. They swim in lines, with the geese (females) in front, the goslings following, and the ganders bringing up the rear.

I realized then that except for the one family that had had a nest on this pond—on the tiny island barely large enough—all of these geese were new here. One family, the one with three young, had come through the woods from the much larger pond by my house about two kilometers away, where, after prolonged battles with many pairs this spring, it was the winner. Another pair and their recently hatched young had walked from another pond still another kilometer farther. Meanwhile, up to thirty nonbreeders had been here all along, aggregated peacefully as a group at this pond until shortly after the various pairs with the young arrived. Now it was no longer a gathering place of the nonbreeders. All except four (who stayed strictly by themselves, separated from the mega-group) are gone. Where have they gone? One of them had been Niko, a tame goose I had raised from a gosling two years ago. She had visited us at our house the week before, but only for two days, and then she, too, had flown away. As far as I could tell, she had not yet lost a single wing primary, whereas it looked to me that these, the breeders, had lost all of theirs.

I knew very little about these geese, but I was intrigued. I know that young are sometimes "adopted," but I am not sure how willingly that occurs.[3] On May 9, when there were only the nonbreeders and three families with young here, I saw one "family" with seven (at least one too many to be all their own). One of these young goslings was attacked mercilessly by a parent of the group. At least four times it grabbed the youngster in the bill and shook it to fling it aside. I was afraid it would get killed. The young gosling came back several times, but finally it ran away and joined another family where it was accepted without even appearing to attract attention. The young of both families were of the same age and size. How had this individual been discriminated against, and why? In a previous year I had once seen a family approach a mega-group, and the two respective ganders had squared off and battled viciously until the ground was littered with feathers.

The current superfamily of five pairs moved around as four "families" that stayed separate but near each other when they swam around in the pond, but aggregated and sometimes mixed together when they foraged. It was plain to see that the geese maintained their family structure, which was presumably maintained by attractions and aversions that likely have adaptive reasons. These highly conspicuous birds are undoubtedly welcome meals to canids, and if such predators could isolate them, they could get them. We humans have over evolutionary time also associated into mega-groups, and most probably for similar reasons of defense and offense. Long-term focus on relatives and associates, potential mates and young, has been as relevant for our survival and reproduction as that of the geese.

As just mentioned, one of the nonbreeders that I had seen a week ago was Niko. About a year earlier, I had seen a one-on-one attraction between her and a wild gander when she was only one year old (the geese do not breed until generally three years of age). Our family had raised her since she hatched from the egg on May 27, 2007, and she had been free around the house where she saw other geese. After spending all summer with only us she finally left us on the evening of September 10 to follow others who had visited and called from our nearby beaver bog. Suddenly she returned one morning to our doorstep about eight months later, on May 11, 2008, and she was not alone.

On the morning when Niko arrived I was awakened from sleep by an emphatic calling of a goose from the bog near our home. The cries were so emphatic and animated that I got up to investigate. It was dawn, around 5:30 AM. I saw two geese in the distance, and one (most likely a male because it was larger) was swimming behind the other, gesticulating and calling wildly, with his head and neck held toward the other at water level. They continued like this for another hour, and I then went back inside the house. Several minutes later the pair of geese suddenly flew up onto our lawn. Now I saw that one of them was Niko, and with her was a big handsome gander, who had been and was still gesticulating to her.

She immediately walked up to me, and he hung back and looked. He held his head very high and seemed on edge. Nevertheless, he stayed. At 9:30 AM he made the to-me-long-familiar head-wagging motions that meant "it's time to leave," and she assented by joining him in giving the same gestures. She then lifted off and flew and he followed her back into the beaver bog.

From this day on Niko and her gander were inseparable. They came every day and would stay for most of the day. We named him Leroy. He, a wild goose, was at first hesitant to come up to our lawn and be near me. Whenever Niko flew in and then walked toward me, he went with her partway. After she had visited with me and snacked on the cracked corn that I offered, she walked back to Leroy, who waited. After awhile he would start nodding his head, indicating it was time to leave, and she also put her head up and soon flew. He launched himself a fraction of a second later and followed her.

After several days Leroy allowed me to come within about ten feet of them. He stood guard while Niko, my young former pet still at least a year from reaching sexual maturity and almost two months past Canada goose nesting and mating time, fed on the cracked corn that I was putting out for them. He always waited until she was finished feeding and put her head up, and only then would he go to the corn and feed. If I didn't know their history, I would not have been able to distinguish them from a mated pair. They were gone after two weeks, and I had no idea what happened to them. I suspected it was a trial love, because surely she could not have been ready to have a mate; she was far too young and the season was far too late.

As I write now, another year has passed, and Niko again arrived, this time on June 5. I saw her standing on the lawn at 7:00 AM. She was alone. Her head was high and she was alert. Not yet knowing who it was, I went out and cautiously walked up to her. She seemed only slightly hesitant, then put her head down and walked toward me. I picked some clover out of the lawn and she took it from my hand. I sat down and she soon stood on one leg, made her soft purring comfort sounds, preened, stretched a wing, put her head on her back and closed her eyes. She was serene and calm. It was as if only a day had passed since I had seen her last, not a year. That evening she flew back to the beaver bog, where until a few days ago there had been a cacophony of goose commotion from the territorial battles that had culminated in the parents of our present gosling, Clover, winning breeding rights. By the next morning Niko was back, and this time she nonchalantly walked in through the open back door of our house and reclined on the floor, leaving two calling cards (green ones) before returning onto the lawn. I had already taken mug shots to confirm her identity (from the details of her white face-chin pattern), but her behavior confirmed who she was in case there had been any doubt. Like Peep, another goose I had before her, I expected she would again leave in the fall and come back next spring with a mate and try to nest here.[4] Yet she left after the second day, near the time when all the other nonbreeders were leaving.

*The physiology of bonding.* Birds exhibit attachment to a place, to a mate, and perhaps to their nest and their young. The love objects are individually specific, and these are the conditions we associate with the psychological phenomenon of concentrated attention we call "love." But attachment to a mate, as such, need not necessarily be by the mechanism of love. There are various other potential means. In deep-sea angler fish monogamy (or polyandry) is self-induced by a male (or several) who physically attaches himself by the mouth into the skin of a female and becomes a parasitic appendage for life. In storks the monogamy is apparently accomplished primarily by the pair's mutual attachment to the nest, where they meet and mate with whomever is there. However, in many birds as in a few mammals, psychological attachment is necessary where the partners are necessarily at times physically separated, and where the individually matched partners need to be identified so as not to be confused with others.

The chemistry of emotional attachment involves oxytocin, a nine-peptide hormone produced by the brain. The hormone binds to other neurons in the mid-brain inducing them to release dopamine, a neurotransmitter that, when combined with specific sensory stimuli, induces pleasurable effects and hence promotes specific behavior and physiology associated with those specific stimuli.

Oxytocin is an ancient hormone that is similar in structure to vasopressin, a hormone that is involved in water balance. The genes of these two hormones are close to each other on the same chromosome and are thought to have been derived from duplication followed by differentiation. It is presumed that the ancient hagfishes, some 500 million years ago, had the vasopressin-like hormone, which was duplicated some 100 million years ago to then evolve in fishes to the oxytocin that promotes sex by reducing fear in ovulating females so males could approach them for fertilization. Ultimately it led to bonding. Virgin female sheep bond to lambs that they would normally reject if given oxytocin, and they reject their own lambs if given a chemical antagonist to oxytocin. Oxytocin is released by the brain in both males and females during sexual activity, and through the axis of inducing pleasure it promotes mutual bonding. In voles, oxytocin promotes social bonding. The monogamous prairie voles are social animals in which the males attend to mate and young. In the solitary montane vole, in contrast, the males are promiscuous and do not attend to their young. These differences in behavior are proximally related to different locations (different neurons) equipped with the oxytocin receptors in the brain—possibly to those sites that are related to identifying individuals—and ultimately influence ability to rear offspring as demanded perhaps by different environments. In the prairie voles the hormone binds to receptors in the midbrain that release dopamine, which reinforces the male's behavior of bonding with both his mate and his offspring, presumably because it induces pleasure.

The capacity for social bonding through pleasure, given an original advantage in rearing offspring, could then be applied to other advantages. Probably these mechanisms apply also in birds, although birds have slightly different structures of these compounds. Instead of oxytocin and vasopressin, they are called mesotocin and vasotocin. The induction of pleasure by linking sensory stimuli through oxytocin that is applied to a mate for monogamy and to young for parenting could possibly apply also to love of home territory or habitat. I suspect it could also shape all sorts of other adaptive behaviors that involve long-term and sustained response where ultimate rewards are neither obvious nor conscious.

In humans oxytocin induces trust, and in concert with the appropriate stimuli from individuals, also bonding. Strong bonding is likely an adaptive trait in humans because it makes cooperation possible and improves competitive advantage.[5] Just as there are human behavioral pathologies where there is little oxytocin secretion or where the receptors for oxytocin on neurons coding or facilitating are missing, it is likely that there are other individuals in whom oxytocin production is "in excess" relative to the general population, and where the receptors for the hormone are distributed in different amounts to different neurons, thus specifying different love objects.

In us "true" love develops where it is reciprocated. In this way the bonding is adaptive because it is not wasted. The process is like a positive feedback loop, where the more you get the more you give. It may potentially even be relevant in someone predisposed to love birds and everything they represent, with the positive feedback coming from what they learn from them.

Curiously, although hormones and their effects are universal, the study of love seems to be restricted to humans, even though there is ample reason to believe that it is a basic adaptive response of many other vertebrate animals. Furthermore, those who study love differentiate and classify various kinds and components of love, just for humans alone.[6] Not everyone agrees on the differentiations. Generally, though, they agree that (human) love involves need, attachment, and compassion. I think popular song writers already knew that in the 1960s, if not earlier. As I have indicated, love is a physiologically complex mechanism of bonding in which natural brain opiates (that are ubiquitous in birds, mammals and other vertebrate animals) play a large role in valuing and focusing. Love and attachment mediating romance by oxytocin and vasopressin can be parsed out to a sex drive influenced by testosterone, norepinephrine, and epinephrine.

Although the role of chemistry in love is presumably universal, its nature and quality and the objects of its focus change through time and the age of the individual and likely vary among species. Love is ultimately an adaptive physiological and psychological mechanism for promoting long-term attachment and commitment where it is required for successful reproduction. Of course it does not require reproduction. It is like an organ, a precondition that makes its use possible. It is thus generally focused primarily and most directly on potential mates and offspring. Until proven otherwise, scientific objectivity demands that, like other physiological processes, it is not a domain exclusive to just one species, but more likely applies to other vertebrate animals as well. We can, however, quibble over its diversity since we can safely presume that it is tailored for specific requirements that vary between situations and species. For example, in humans the argument goes that love evolved to induce men into bonding, since they cannot predict (contrary to polygynous apes and monkeys) the female's time of ovulation. The lack of sign therefore required males to be with females continuously in order to mate and to jealously guard them to prevent cuckoldry.

Obviously this adaptive argument, although it may apply as a significant variable in human evolution, provides less explanatory power to birds, who apparently are well apprised of the approximate time of ovulation in their mates, since even in monogamous species mating is restricted to only the very narrow time window near egg laying. In most species there is no mating at all during about 97 percent of the year. In the polygynous species, on the other hand, there is no attraction to any one specific individual in either sex, and the brain opiates and sex-drive component of the love phenomenon are focused instead on very specific physical attributes that are unrelated to individuals, as such, and in them mating occurs over a period of months in some species, provided fertile individuals are available.

One of the strong interests that we have in other animals (as we also have in one another) is not just what they do. We also want to know why they (and we) do it. For that we need to learn what drives them to behave as they do. However, despite our own huge emphasis on feelings or emotions in trying to understand ourselves, the topic has been practically taboo in the considerations of other animals, for whom instead we categorically presume that we cannot know what they feel. However, a more conservative approach is that feelings of love and attachment, despite their imponderable qualities, are universal attributes of other vertebrate animals, just as our physical construction is virtually identical to theirs, except for the minor exterior adaptations to our differing physical environments. By invoking Occam's razor it would be a greater error to presume that animals are not driven by feelings than to suppose that they are. Instead, currently we seem to have decided that humans are exempt from the rules that obtain across the rest of nature, and that sophisticated emotions are our exclusive province.

Although Occam's razor does not provide proof, it does narrow the search and tells us where to begin looking. I believe we are now at the brink of new discoveries because of new methods. We can now almost literally look into the brain, and functional magnetic resonance imaging (fMRI) is showing us that specific areas of our brains light up (become active) in correlation to activities, thoughts, drives, sensations, and emotions. We know, for example, that the sex drive, the induction of the emotion of romantic love, and that of maternal love each light up very specific (and also overlapping) areas of the right ventral tegmental area and the right postero-dorsal body of the caudate nucleus of the brain stem.[7] These centers are involved in the dopamine-dependent reward and motivation neural pathways. With respect to love, which involves attraction to a specific individual, they would function to reward focus on the individual, and that focus presumably evolved when a prolonged attachment to that individual was required for reproduction. (Grief activates other brain areas, but that from loss of a loved one reminds the subject of the beloved and activates the reward center associated with that loved one.[8])

Birds' brains are organized differently from ours. However, like us they have experienced convergent evolution for likes, dislikes, attractions, and bonding. Therefore it is reasonable if not certain that they also have particular brain areas allocated that are active in response to different situations that would predictably correlate with analogous behavior and situations that would support such behavior. It would be surprising indeed if fMRI of them would not reveal similar predictable patterns such as those found in mammals, as well as showing likely differences. (The main problem, however, would be that it would be more difficult to "ask" them what they are feeling; when held down to be measured they might feel mostly stress.)

It is easy to put ourselves into a special category simply because we do not intimately know other animals. Our lives are too distant from others to be able to make valid inferences (except possibly with respect to our dogs, with whom we live and mutually interact for years). My shift in the way I looked at birds occurred when I raised "Peep," a Canada goose hatchling, to adulthood and then followed her life as she mated with a wild gander and nested in our beaver bog. Through her and her mate, and later their associates, I was introduced to the world of other geese in the wild.[9] Nevertheless, I was of course never really *in* their lives. I was still only an observer—trying to understand them objectively yet beyond their surface reality.

Subsequently to and probably because of my geese project I had the good fortune of getting to see some details of the inner lives of Canada geese from an unexpected angle—a woman who became a surrogate goose. She did not "study" the geese, but literally became so closely integrated with them that she "became" part of their world and vice versa. Ruth O'Leary, an elderly "goose lady" from Priest River, Idaho, lives in the country with her son a half-kilometer from the nearest neighbor. She has four other grown children and had lost her husband after caring for him through his ordeal of dying from cancer. I tell her story here in greatly abbreviated form, drawing mostly from her words. Her life with adopted geese has been and is, she told me, "the most amazing experience." She talks freely of "falling in love" with one or another of her geese and told me, "I find myself looking out the window often expecting to see Little Sir Laughsalot and Lady Diana" return in the spring. She speaks of them in the same way as she speaks of her former husband and son. Her best friend had also raised a wild goose and was with it continually for four years before she was seriously injured in a car accident. She had to leave the goose to a caretaker. When she returned from the hospital two weeks later, her goose was so happy he leaped into her lap, honking and honking and rubbing his bill on her face, and she was so happy to see him that she started to cry. O'Leary muses, "If there is a goose heaven, then that's where I want to go when I die." Had I heard these people-geese stories prior to my own experiences and studies of the same species, I would probably have dismissed them as fiction. However, on almost every point of fact I found concurrence.

I begin here with Ruth O'Leary working in her flower garden on a day in May 2003. She had a Canada goose gosling beside her. She guessed that the baby was only two or three days old because it still had its egg tooth at the tip of its bill. A neighbor, who had been walking her dog, had heard it crying, found it running alone in the weeds, and assumed it had been abandoned. It may have been, but many such apparent "orphans" are not orphans at all. Knowing Ruth had raised a Canada goose, named "TB" for "Tinker Belle," the neighbor had brought the little gosling over, and Ruth had held it in her arms as she had the baby TB two years earlier. After being held, the newly adopted foundling immediately stopped crying. After that it was with her "every minute, day and night, until that fall when it learned to fly." This gosling came to be called LS (Little Sir). Years later it took a mate who was named LD (Lady Diana).

While Ruth was weeding her garden on this May day in 2003

with her still-tiny new foundling LS, two adult geese made an unex-pected appearance. Honking loudly as they flew overhead, they cir-cled closer and then landed nearby. One of them came over to Ruth. It was her old friend Tinker Belle, the previous adoptee from two years ago.

Ruth had been a parent to Tinker Belle until the year before. Until then TB had always been by Ruth's side, day and night. Never-theless, TB had eventually found a mate, even though she stayed very much attached to Ruth. She would still come inside the house and sleep with Ruth as she had since a gosling. In the daytime she then stayed with her mate who came daily, and when he left in the evening she was restless. One night TB "purred softly and rubbed her bill across my face and hair. I knew she was planning to leave." And the next day TB did indeed leave with her mate, after circling again and again. "Eventually she took off and her honking was like a desperate cry as she left with him." Ruth had tears running down her face, thinking she would never see her Tinker Belle again. Seeing her now return with her mate a year later, Ruth was jubilant.

The gander hung back, but Tinker Belle walked up to Ruth as though she had never left and stepped onto her lap, joining the new baby, Little Sir, that she had just adopted. She ate the dandelion greens that Ruth offered, and after the snack TB followed Ruth into the house and inspected all the rooms. She was apparently looking for a place to put a nest because she started pulling the covers off the bed (where she and Ruth used to sleep) with her bill. She also went into the living room, where she used to watch television with Ruth two years before. She pulled out a tape, which fell to the floor, then looked up at the television. Ruth took the correct tape, *Fly Away Home,* off the shelf and put it into the videocassette recorder. Tinker Belle leaped up onto the couch and watched more than half the movie— one she had watched often before. Eventually Tinker Belle walked back outside to rejoin her mate, and at dusk, when he began bobbing his head to signal his intention to leave, she took off first and he followed.

After this visit TB and her mate came back every morning and left again in the evening. All went well, until one day her mate was missing. TB circled the valley again and again, calling loudly in a frantic voice. That night she again slept with Ruth, now also with Little Sir on her bed. After three days TB gave up looking for her mate and sat with her bill under her wing. She refused to eat, didn't react, and became "inconsolable." I speculate that grief represents a perception of loss, which motivates a strong desire to regain what is lost. Perhaps such feelings are not possible without first having a need, a prerequisite for love.

A week went by. Tinker Belle became weak and staggered. Ruth inferred that "her grief was profound." For three days Ruth, TB, and Little Sir ate, slept, swam, and bathed together, and Ruth believed TB eventually got over her depression because of the gosling. Tinker Belle later flew to a big neighboring pond where she joined a flock of geese, and Ruth didn't see her again all summer. That fall a large flock of geese was flying over, and as Ruth looked up she saw one goose break away from the flock, circle the house, and then rejoin the flock and fly on with it. Ruth presumed it was a gesture from her Tinker Belle, the goose who had been her constant companion and who had slept with her on the same bed every night for two years.

In the meantime, Little Sir had grown into a magnificent gander. He always stayed close to the house, where Ruth had a small pond at which she fed both her domestic and wild geese. Like TB, LS returned to the house every night to sleep on Ruth's large lounge chair in the kitchen, which served as their bed. He slept on Ruth's chest wrapped in his special towel, which Ruth washed each day. When he awoke in the morning he nibbled on her chin, telling her to get up and feed him his morning liquid corn gruel. In the evening when she took a shower, LS insisted on being included, and afterward he swam in the tub. When they were outside he would scurry to her side and hide under her shirt if a hawk flew by. He was alert to strange noises at night and would tug at her sleeve if he was scared. As he grew bigger Ruth would run and flap her arms to try to get him to fly. After the first time he was airborne, he circled back and landed by her feet, nodding his head up and down as if he were laughing. Thereafter he was known as "Little Sir Laughsalot." LS became excited by the sound of any engine; whenever she started the lawnmower he came running and walked between her and the mower. He also loved to ride in the car, and she drove with him everywhere. He followed any vehicle, and "the louder the engine, the quicker he was to fly after the vehicle." (I found the same spontaneous, strong attraction of my tame goose Peep to my truck when I was in it and she heard the engine running. To her, the noise of the engine was apparently instinctively recognized as equivalent to the noise of a flock that she had to follow.)

Near his second birthday, when he was still unmated, one of Ruth's domestic geese, Annie, started laying eggs. LS stood and watched. After Annie had laid her egg she carefully covered it with soft down and grass and then came out of the goose shed and went to the pool. Little Sir walked into the goose house and to the nest and gently removed the down off the egg, then sat on the nest a few minutes before stepping off again and carefully replacing the grass and down. He joined Annie at the pool and later that day mated with her. Neither then nor later did he show any attachment for her, but he apparently detected her fertile condition and did not miss the opportunity to mate with her.

When LS was only a year old a young wild female, his eventual mate, started coming into the backyard every morning. Both grazed and appeared to pay no attention to each other, but after two months this female began to call when she arrived, and LS answered her. After awhile he also followed her. They went into the neighbor's field where they fed together, and when she flew off he would follow briefly but then come back. The next spring this same female was one of the first geese to arrive, and now when she honked LS answered excitedly. Again they wandered around together. Curiously, LS now started attacking people.

The following March, in 2007, after his female friend had been gone since fall, LS woke Ruth with loud honking, leaped off their bed, and ran to the back door. Ruth opened the door, he rushed out, and there was his friend. They spent the day close together—feeding, swimming, and napping. Ruth summarized, "It had taken three years, but the little lady had finally stolen his heart." She said he was so enamored with her that "he couldn't contain himself." Ruth named her "Lady Diana" because "It seemed to fit her quiet, gentle personality."

For four years LS had been with Ruth "constantly day and night." He had never flown south in that time. However, he liked to fly or engage in the equivalent: riding in the car. Every day in the winter Ruth took him for a ride in her car, and she always preceded this treat by saying: "Baby ride?" When Ruth opened the door he ran over and eagerly jumped in, standing on his mat on the front seat, and as she drove he stood tall and looked out. Eventually he would show excitement and clear understanding when she asked him because he would run to the car and be ready to jump in as she opened the door. Sometimes in the winter he circled the valley where they live and then flew off. Ruth would drive after him to try to find him. When she caught up with him, she would stop and call. He would come down, and as she said "Want ride?" he would jump in.

Little Sir continued to be extremely gentle to Ruth after he had fallen for Lady Diana, and he still slept with her at night. He sat on her chest, rubbed his bill across her face and purred. For three weeks he spent nights with Ruth and days with his Lady Diana. They greeted each other every morning with a display of affection that lasted several minutes. He was, however, still attached to Ruth, and groaned "Uuh, uuh, uuh" with his tongue hanging out as he would drape his neck over hers. Ruth says, "The last night Little Sir spent with me he was even more affectionate than usual. He sat on my chest, rubbed his bill across my face and hair, and purred. When I dozed off and my arms slipped off him, he tugged at my sleeve to get me to put my arms back around him. I stroked him and told him I loved him. I knew he was spending his last night with me."

After their usual affectionate display when LD arrived the next morning, LD and LS began looking for a nest site. From now on he no longer spent his nights with Ruth, but with LD instead. That morning Lady Diana flew toward a nearby wooded area, and he followed. But he came back to the yard and called LD, who joined him. Together they searched the yard—the raspberry bushes, around the shed, underneath a trailer, and then LS flew onto a big nest box that Ruth had specially installed as a roomy, secure place that would be safe from coyotes. But LD didn't join him—instead she flew back to the woods. This time he didn't follow her. Instead he stayed on the nest box and continued to call. After she didn't come he flew to the woods and immediately flew back. This time she followed and looked at the nest box, but only briefly. Then they grazed and napped, and while they napped Ruth refurbished the nest box with fresh hay.

Early the next morning they both returned to the yard, and LD again flew toward the woods, and LS followed. But within a few minutes he flew back, and she followed him back, landing on the nest box. They called to each other, over and over. He refused to leave the nest box. Finally she came over, and this time she was apparently impressed with it because she started to arrange the hay, pulling some under and around her. She had finally accepted his choice.

Each morning after this LS and LD flew to the yard and LD would fly to the nest box to lay an egg while he waited nearby. Each morning Ruth took a dish of corn out to him. Instead of feeling grateful, however, he would now lower his head and try to bite Ruth, his friend, parent, and caregiver. But he always first looked to see if his wife was watching. She was. Each morning he became more aggressive. Usually Ruth resorted to retreating to the back door, and he stayed and slurped down his daily ration of corn mush. But one day shortly before nesting began as Ruth hurried back to the house, he didn't even stop to eat. He ran after her, drew back his wing, and hit her with force that, in Ruth's words, "I could not imagine him capable of." She sat down in pain, saying, "Ouch. Hurt." And he looked her straight in the eyes. "He stood for a minute just looking at me, then slowly turned and walked back to his lady. He turned back around and looked at me sitting on the ground." Ruth then hobbled back into the house. By evening her ankle and entire foot were blue. It was several weeks before she could walk without a stick for support.

Was Little Sir, Ruth suggested, trying to impress his loyalty to his mate by giving proof that she had nothing to fear of his previous bond with Ruth? Before and while he was running at her, he would stop to look back at his mate as if to make sure she was watching. My own interpretation, based on my field studies in Vermont, is that LS was mate-guarding. According to goose "logic" anyone in proximity to his female might also cuckold him, so an alternate hypothesis is that his response was based on jealousy, to discourage Ruth from and possibly punish her for getting near his mate. During this nesting phase he also viciously attacked wild geese that came near. Little Sir guarded the nest full time, never taking a break until his mate took her short break every afternoon to bathe and feed. She then flew to the pond, and he followed her. Often he sat in the nest box with her, and in the last three days before the young hatched and departed the nest, he never left her side. The young hatched on May 19, and two days later Lady Diana flew to the ground and began to call to the goslings from there, apparently to get them to jump the six feet to the ground. Little Sir stayed in the nest and occasionally nudged them with his bill. Lady Diana then flew to a pond a mile away while Little Sir remained with the goslings and repeatedly called. She returned forty-five minutes later to lead the whole brood away, with Little Sir bringing up the rear. Ruth followed along behind, just to make sure they made it safely through the woods to the big pond. There at the pond she saw a pair of geese with twenty-five goslings, but from then on and through the rest of the summer she lost contact with her goose family.

One day in early September she heard honking, and recognizing the excitement she ran out of the house—and there they were, Little Sir and Lady Diana, and they had brought their five "teenage" youngsters. Ruth ran to meet Little Sir and called, "My baby came back. My baby! My sweetheart! Do you want some juice?" (his favorite—warmed-up frozen corn). He greeted her with his head bobbing while Lady Diana remained at a discreet ten or twelve feet away. He came to the house with her and yanked at her pants while she warmed his corn, as he was getting impatient.

After that the family flew in every morning at 8:30AM. At 10:30 they left to fly back to the big pond, and at 3:30PM they returned and stayed till 8:00, then returned to the pond for the night. "It was as if they watched the clock, so regular were they to arrive and leave." Lady Diana would always be the one to signal intent to fly. The young would follow her, and Little Sir took up the rear as the family made their daily tour.

Finally, one day in late September, after they had not shown up for some days, Ruth saw the first V's of geese pass overhead. Little Sir dropped out of a passing flock and landed at her feet as the flock flew on. Had the family gone on without him? The next day Ruth and her son went over to the big pond, where many geese were aggregated. She called "Baby! Baby!" and Little Sir came swimming over to the edge of the pond, stepped out, and nodded his head in greeting. He later rejoined the group there, and that was the last she saw of them that year.

Six months later, at the end of March 2008, Little Sir flew in, alone. But he spent only an hour before he looked off in a northerly direction, gave the sign to leave, and lifted off. The next day he was back, but this time his wife, Lady Diana, was with him. The couple again nested in their old nest box. Little Sir was again very aggressive, both to Tinker Belle, who had also returned (but who had apparently lost her mate), and to Annie, the tame goose with whom he had tried to mate before he wedded Lady Diana. He attacked other geese viciously, beating them with his wings. He attacked any group of geese approaching their pond, trying to land on it. Even before they had a chance to land he was already ripping chunks of feathers out of them. Ruth said, "He was a terror." Once when he attacked Annie, he stood on her until Ruth rescued her. He attacked another tame goose and broke her wing at the shoulder so that it had to be amputated. He made two exceptions: his favorite daughter came to visit with several others (possibly siblings), and he seemed overjoyed to meet them. Also, he never again attacked Ruth. Perhaps he had come to realize that he had no grounds for jealousy; which boils down to fear of cuckoldry in a monogamous relationship.

The next spring, when Ruth had known LS for several years, he returned with his old mate LD. Oddly, he did not care to be with Ruth any longer. He always honked when he came so that Ruth came out to meet him, and then he would turn his back on her "as though he didn't want anything to do with me," Ruth surmised. The couple nested nearby and had five young that they led away, but they came back to the old premises by Ruth's house after a few days, and they were then without the young. Meanwhile, Tinker Belle came back single, having lost by now two mates. Unlike Little Sir, although she also came every day, she always flew in silently, without honking, and she didn't want to have anything more to do with Ruth, either. Her geese had finally reached independence. It had been a fabulous experience.

Very few of us will ever have the opportunity to get to know a wild bird as well as Ruth did, or vice versa. We would probably never have the time or the tolerance. However, if this can be an example, it shows we identify with geese because we can empathize emotionally with their bonding behavior—a behavior conspicuous both between offspring and parents and between mates. We also understand their method of childrearing, which requires a monogamous relationship with cooperation and trust. We generally assume such mutual identification to explain bonds between mammals (such as between us and our dogs), but the mechanisms of bonding are basic, that is, primitive, and hence they are likely adaptive and serve similar functions relating to need and survival.

2

———

MONOGAMY

*JUNE 10, 2007.* I'M IN VERMONT AT MY desk with a window facing a grassy slope dotted with a scattering of blueberry bushes and fruit trees. It is a warm, sunny day. Tiger swallowtail butterflies sail by. Several days ago they fed on wild honeysuckle, whose bloom is now past. In a few days the butterflies, too, will be past. The plants, insects, and birds are all on tight and interlocking schedules. I had put up bird boxes (with loose covers so I can look in) several years ago, and this year there is again, as almost every year, a pair of tree swallows in one of the same boxes. Four eggs hatched two days ago.

Yesterday one of the swallows dive-bombed me repeatedly when I neared their nest. After each pass it flew up to about fifty or sixty feet to turn around and dive back down at me like an arrow. Within a foot or two of my head it chattered loudly in a rapid series of clicking sounds and then veered off within inches of my face. It made one pass after another until I left. These swallows usually do not attack people. This morning one of the swallows lands at the nest entrance and flutters there while peeking in at intervals but unwilling to enter. A second bird with a more shiny head flies in from over the nearby beaver bog and slips in through the nest entrance without the slightest hesitation. After it leaves, the flutterer returns to the entrance, flutters there again while peeking in at intervals. It flies off, making a loop around the field and the bog, only to come back to the nest box and again repeat the same behavior. It stays for minutes at the entrance, sticking its head in and out at intervals but not entering. I'm baffled and leave my desk to walk to the nest box to see what I can find out. "She," the flutterer, totally ignores me even as my head comes within a foot of hers while she now perches directly on top of the nest box and preens herself. She casually nabs black flies whirling around my head and hers. Finally she lifts off, circles, and flies past me. She comes back again, treating me as though I were a post. Dozens of times she does this, and as always after each pass she lands at the nest-box entrance and repeats her fluttering and poking her head in. I lift the cover and look into the box—all four pink babies look fine.

I re-enter my office and continue to watch from there. She again continues the flights and the flutterings, but in the meantime another swallow now slips unhesitatingly through the entrance in and out of the nest box. This bird sometimes remains inside for ten minutes at a time, then finally comes back up to the entrance from the inside and sticks its head out the entrance hole. It stays perched there for over a minute, continually inspecting the ground below and all around to the sides. Eventually it leaves, and at that moment a third swallow comes and slips in but leaves within a few seconds. Within a minute the first bird makes the repeated passes over the nest box and then, when finally stopping at the nest entrance, just flutters there.

In the afternoon the presumed female was again alternately perched on the nest box and fluttering in front of the entrance as though trying to get in. I again went to "her" and she was as tolerant of me as before. I enjoyed the intimacy with her—she was close enough that I could have reached out and touched her, and my nose was practically at the nest entrance. Two swallows then came swooping in and both dive-bombed me. The "flutterer" joined the pair in circling me, but not in the dive-bombing. Oddly, during the hours that I had watched them none of the three ever displayed the slightest sign of intolerance toward each other, and the pair successfully fed (and eventually reared) their brood of four young. Tree swallows usually fight fiercely for nest box ownership—I have seen fights at nest boxes where the birds grappled and fell out of the air and continued their tussle on the ground. I found one pair so engaged in their fight that I was able to walk up to them and pick them up from the ground before they untangled in my hand and flew their separate ways.

The next spring, in 2008, three swallows again showed up. I saw them first in the nearby beaver bog, and I was apparently remembered because I was immediately attacked by one of them, even though nesting had not yet begun. But that year only two birds ended up at the nest box. I learned that the tree swallows dive-bombed someone (me) who had opened their nest box to look inside, but they ignored other people who had not done so; they recognize individuals. They have the capacity to have individual friends and differentiate enemies of their own and other kinds, and apparently they also possess a long memory. These attributes are prerequisites of monogamy, which is based on psychological bonding to a specific individual.

Like many other people in the north, I stock a birdfeeder with sunflower seeds near our house. It is visited by black-capped chickadees, blue jays, tufted titmice, and sometimes by one to several evening grosbeaks, purple finches, goldfinches, redpolls, pine siskins, cardinals, and downy and hairy woodpeckers. I see no sign that any of them are bonded to each other. Yet all are typically classified as "monogamous" species, as are approximately 90 percent of birds.

Another species of the same general group of passerine songbirds is also classified as monogamous, but it acts quite differently. My second bird feeder, at the edge of the garden, is a log platform about three meters off the ground (high enough to keep our dog away) where I provide meat scraps for our local raven pair. All of the other birds, except for the finches, on occasion feed there also, but the raven pair comes daily. These birds behave differently than all of the other birds; they stay together almost constantly as a couple. Every morning at first light the pair fly here from the cliff a half-mile down the road where they nest and where they sleep at night. Long before they nest in March and long after they finish in the fall they still perch closely side by side, high on a limb of the great ash tree by the house. They preen each other, making soft grunting comfort sounds, and seem to be enjoying each other's company before eventually gliding down to the platform to feed. They perch side by side on a branch in falling snow, and they gently tweak the feathers on each other's heads. Their conversations say nothing that is relevant and meaningful to any of us, but to each of them it is undoubtedly meaningful. After they schmooze they both fly off together. They never quarrel at the food, but if a third raven should come anywhere near, then both instantly give vigorous chase and drive it away.

Raven couples are easy to spot. About twenty years ago John Marzluff and I captured a crowd of forty-two ravens and put them into a huge outdoor aviary, and in this crowd we instantly recognized a bonded pair. The two sat next to each other continually, preened each other, and "talked" to each other by gentle and soft vocalizations. Their devotion had little to do with breeding season, as such, when most other monogamous birds pair up. When I have released unpaired ravens from the aviary, they flew away immediately. But if I released only one of a pair, it did not fly off and leave the other behind.

The strong bonding (and apparent hating) that ravens display toward specific individuals may be at one end of a long continuum in birds. Like any other "extreme" morphological, physiological, or behavioral attribute, it must have evolved as an adaptation, but it is not clear how the raven's monogamy differs from other kinds of monogamy and how it has helped their survival and reproduction. Ravens are long lived and stay permanently in an established territory, and after proving the effectiveness of their partnership by nesting successfully, presumably each member of the pair benefits more by staying with its partner than by switching.

Many of the so-called monogamous birds regularly mate outside their pair bonds.[1] They are clearly not, by inclination and fact, monogamous—at least not by the traditional definition, having only one mate. If the male gains paternity with females other than his own, it is called polygyny, or polyandry if the female procures matings by soliciting other males besides her mate.[2]

Males may nest with only one female although mating with several, and they could take another female for a second nesting at the same time or in the next breeding attempt. Thus, they may be serially and socially monogamous but not necessarily sexually monogamous, and they may at the same time be considered polygynous either within the same or different seasons. Furthermore, such definitions become even more clouded when one considers that the cuckolding is achieved not only by the distribution of sperm by the males, but also by the distribution of eggs by the females into the nest of another male.

Pairwise parenting is rare in mammals, insects, crustaceans, fish, reptiles, and amphibians, although it occurs sporadically in all of them.[3] It is common only in birds. Birds routinely team up into one-on-one male-female partnerships, and such pairs are so obvious and conspicuous that we take monogamy almost for granted. Not being sure what "monogamy" means, since many birds also routinely have extra-pair matings, I consulted the *American College Dictionary*, which defined the term as: "Marriage of one woman with one man" and "The practice of marrying only once during life." I didn't know for sure what is meant by "marriage" (except that it is a social ceremony, lifestyle, legal document, or moral commitment, and not necessarily a biological term), but that didn't matter because the more relevant, specifically "zoological" definition of monogamy was: "The habit of having only one mate." If that refers to the practice of having only one mate during a lifetime, then few birds would be monogamous. We are conditioned to think of them in this way, however, so we hugely broaden their monogamy by qualifying it with the adjective "socially," and by deleting the restriction on the term we apply to ourselves, namely "once during life." This much broader zoological definition nevertheless clearly differentiates between a male mating with more than one female during any one breeding season, which is then understood to be polygynous, and a polyandrous female that mates with more than one male in a breeding cycle. In questioning the contention that 90 percent of birds are "monogamous," John Alcock in his textbook *Animal Behavior* adheres to the zoological definition, which can then be used to make a distinction between social monogamy and genetic or copulatory monogamy.[4] But there are even more complications in trying to pigeonhole species into definitions.

A species that is defined as "polygynous," given the standard near-even sex ratio, has by definition at least as many males who likely have only *one* mate, if any. And in that same "polygynous" species, in which only a small minority of males have many mates, most if not all of the females may, in fact, have only one mate. That is, they are monogamous. So it seems to me that species definitions don't count, unless one has a gender bias. Conversely, most of the so-called "monogamous" species are often widely dispersed in space and thus have little or no opportunity to mate with others; if one were to experiment one might find that they would mate with more than one female, or one male. There may indeed be natural experiments. In one study by Toni DeSanto, Mary Willson, Kristen Bartecchi, and Josh Weinstein of nesting sites of winter wrens in southeastern Alaska, it was found that nests near ground level were much more likely to be attacked by predators than those nests in hanging moss from tree branches, and that affected the wrens' social-sexual mores.[5] The tree branches with moss were extremely abundant and uniformly distributed, which probably made it difficult for predators (primarily red squirrels) to find them. The clumping of food and nest sites elsewhere, however, affected the breeding system. At clumped resources, such as along stream banks, a large percentage of the male wrens had several females (they were polygynous), whereas in the uniform forest environment they were monogamous.

Mating and pair choice are not tightly coupled in most birds. In so-called polygynous and polyandrous species there is no apparent male-female bonding. There is instead relatively indiscriminate mating on the part of one sex coupled with often intense discriminating on the other. At the other extreme are apparently permanently bonded couples who mate only in a very narrow time window (at or slightly before the time of egg laying), during which they may also mate with neighbors, depending on opportunity. In many birds who have successive broods through a season, such as song sparrows, for example, a female may be monogamous in the first brood but mate with several males (probably carefully chosen, that is, not "promiscuous") in the second, even as she remains with the same original social mate to rear their clutch.[6]

One way around the conundrums of definitions has been to add even more qualifiers or adjectives than just "social." Another solution has been to create definitions to refer not to mating but instead to the effort expended in child-rearing, leading to designations of "facultative" versus "obligate" monogamy.[7] In the case of obligate monogamy, two parents are required to rear the young, whereas in facultative monogamy, the male stays around only long enough to ensure paternity. A more workable alternative to the "obligate" definition could be "evolved monogamy."[8] Placing the emphasis on "evolved" is instructive because it makes monogamy an adaptation and releases us from the immediacy of individual proximate causation.[9] In an evolved monogamy we would expect to find pair bonding as part of the reproductive mechanism, in which the reproductive success of each depends on the other.

Few bird species form a lifelong attachment to their mates, and even fewer species mate only with the one to whom they are bonded. So bird monogamy, broadly defined, could thus include "social," "serial," "sometime," and "strict" in addition to "facultative," "obligate," and "evolved." "Strict" monogamy can arguably be found in a parasitic worm, *Schistosoma mansonii*, where the male and female entwine in one's liver in an incessant lifelong copulation and almost ceaseless outpouring of eggs. Strict monogamy is also achieved in many insects where the individuals live only a few days as sexual adults. They mate only once in that lifetime because their lives are so short that they experience only one reproductive cycle that requires just one mating. Similarly, many amphibians and fish are monogamous for the same reasons.

Monogamy is ultimately relative, and the relevant question is this: What is it good for? What are the advantages that have provided the selective pressure to evolve it? The default idea from a superficial reading of evolutionary theory (promulgated by males mostly) is that males are in control and should not restrict themselves to only one mate. They "should" (that is, have in evolutionary history) father as many offspring as possible, and one might predict therefore that they would mate and then leave a female stuck with the eggs and young and go on inseminating the next one. So then why do some birds have only one social mate at a time whereas others have many mates?

The winter wrens provide evidence for one key idea in the evolution of monogamy: How plentiful are resources critical for reproduction and how are they distributed? There are many examples to choose from. A study by F. R. Gehlbach spanning four decades of over a thousand banded screech owls shows that in them lifelong monogamy is the rule, but polygyny occurs when food is plentiful and nest density is high because a male can simultaneously provision more than one nest.[10] In general, females are limited by resources—they require large amounts of food to produce eggs and rear the young. Sperm, on the other hand, are cheap and plentiful. When food is clumped in time and space, females "should" tend to aggregate near it and thus breed together at the same time.[11] These conditions should then cause males to gather nearby and to compete against each other, and the winners would then monopolize the matings of the aggregated females. These males would then be polygynists. In contrast, when resources are widely distributed in time and space, the females will necessarily be scattered, and they cannot then be monopolized. Furthermore, they may benefit from help with child-rearing. Under these conditions a male's reproductive potential is increased by staying with a female he finds and helping her to rear his offspring, which would be less likely to survive without his help. Convergence in mating syndromes between widely divergent species from different animal groups supports this scenario, as does divergence in related species, such as for example the exceptions among the largely polygynous birds of paradise in Papua, New Guinea.[12]

Most bird-of-paradise species feed on fruit that are locally abundant at certain times. The females then breed in phase with these food bonanzas, and they can rear their young alone. In contrast to most other birds of paradise, the manucode feeds on figs that are spaced widely both in time and place, and it takes two parents to keep hunting for the highly patchy and ephemeral fruit-bearing trees. Two parents are required to provision the nestlings, and so this species is monogamous. But as always, this generalization applies only when all conditions are equal, and the genetic history of a group is often deeply ingrained. For example, corvids are highly monogamous regardless of varying lifestyles and resource use. In the same environment in Arizona, scrub jays and pinyon jays have two wildly contrasting solutions for making a living.[13] The pinyon jays are specialists on pinyon pine seeds that are very patchy in time and space, but they are social birds and excellent flyers and collectively able to search out and reach the shifting resources. Scrub jays, on the other hand, are poor flyers. They are generalists and territorial rather than social. Yet both species are monogamous.

Although resource distribution has explanatory power in differentiating many polygamous from monogamous mating systems, it is only one of many possibilities. The "resource" may be mates themselves. Monogamy can assure having a mate and being ready to nest in time in a seasonal environment, but it can be a cost if there is much mortality on migration. In these cases it is better to take advantage of the first available mate instead of waiting for the sometimes unlikely possibility of the return of a previous partner, especially since there is often no obvious benefit to fidelity so long as a mate is available.[14] Bonding strength should thus reflect these differences, and pair bonds may be extremely variable, even in the same species. Laysan ducks, for example, may change pair bonds within and between breeding seasons or extend over nine breeding seasons.[15]

Examples of "extreme" monogamy show a variety of considerations and selective pressures that should facilitate the evolution of

monogamy. The bottom line is still that males "should" father, and females mother, as many offspring as possible. (Obviously no moral imperative is implied; "should" is merely a shorthand way of referring
to a historical pattern.) Under many conditions faced by birds over
the course of evolutionary history, if the male did leave his mate, then
his young would not survive, and it is not just because of the food resources around which females may or may not choose to clump. So,
it's all about the conditions. It's all about the details, and that's what
makes the mating game of birds one of the hottest topics in biology
since Charles Darwin spawned the pillar of biology, the organizing
principle of natural selection, to explain the similarities and differences between the diversity of organisms.

One of the first differences between monogamy as we perceive
it (and perhaps anthropomorphically value it) and as it is in some
birds is that it may or may not have much to do with mate choice. For
example, European white storks (and presumably many birds of prey)
breed with the same mate year after year, and they appear to be the
paragons of monogamy. But what looks like mate choice is really
nest-site fidelity, not mate fidelity. Male white storks, for example, accept the first partner to arrive at the nest at the beginning of the
breeding season.[16] Since older birds arrive first, the same mates often
meet at the same nest for more than one breeding season. However,
if another female comes first to his nest, then he accepts her instead.
There is no evidence that he identifies the female as an individual, and
hence there is no evidence that he is bonded in what we might define
as "love" to a mate. Instead, his love is of the nest site, and monogamy
is incidental.

The commonality of monogamy is that it is a mechanism for
rearing offspring through the cooperation of a pair. The division of
labor required for teamwork in parenting that has then generated
monogamy originates from a diversity of environments where a variety of factors are challenging and therefore play a deciding role for
a mate to provide aid through cooperation and teamwork. As I will

show later, the demands of parenting would otherwise be so extreme (primarily because of predator pressure) that a male could achieve his maximum reproductive potential only by forgoing mating opportunities with other females. He must "put all his sperm in one basket" because that basket requires all or much of his undivided maintenance and commitment. For that, mutual compatibility promotes better cooperation and implies enhanced reproductive potential, and that compatibility is not found with all females and must be developed as well with those that may be available. However, while monogamy may be an optimum, with the first partner it may represent the best of a bad choice if a better provider can be found. Even the simple effort of incubation could be grounds for divorce. For example, most birds have specific schedules of "changing the guard"—incubation duty versus time for foraging. Suppose one partner stays away to forage for six hours per day when the other gets hungry and wants to stop incubating to go feed after three hours.

There may be conflict for incubation as well. In the rock sandpiper, for example, there is great variation in the allocation of incubation time in males versus females. In some pairs one parent assumes all the incubation and in other pairs it is shared. The partners have to "agree" on a schedule and accommodate each other's needs. Possible conflict may be reduced by specific display ceremonies associated with the changeover as one parent comes and the other has to leave. Although we know little about the details of such changeovers, in one case a male was seen to peck his mate on the back while giving chase to get the female to take flight so he could incubate.[17] In this species those individuals that return after migration also mate with the same partner year after year, raising the intriguing possibility that individual behavioral similarities or differences play a role in creating, and then maintaining, the monogamous pairings.

In a species in which one sex does all the incubation, compatibility could involve feeding the mate often enough and with the right food. Cooperation could involve nest building, nest defense, or the amount of foraging effort for the young. All of these factors could vary not only with environment but also with different partners, so that reproductive output could be a function of compatibility. With experience, compatibility should improve and thus facilitate social monogamy.

In another recently published study, Matthew Johnson and Jeffrey Walters from Virginia Polytechnical Institute and State University followed the breeding biology of a population of western sandpipers in a fifty-hectare upland tundra plot in the Yukon Delta of Alaska.[18] This sandpiper is both socially and genetically monogamous with biparental care of eggs and young. Over the eight-year time span of this study Johnson and Walters individually marked (with leg rings) 533 birds to try to tease apart various factors that contribute to breeding success or failure. The researchers found a low return rate to the home territory, which is typical of shorebirds. Many of the birds either died or dispersed to some other breeding spot. Therefore, true monogamy would have been costly. When both members of a pair returned, they reunited 4–26 percent of the time and took new mates 10–29 percent of the time. Both mates are needed to bring off their clutch of four young. If either parent is removed, the nest is abandoned. Mate fidelity, where possible, indirectly affected nesting success. On average 75 percent of nests were lost to predation, but the couples that stayed together bred sooner, had earlier nesting, and suffered a lower predation rate than those who nested later. Apparently, either breeding in a familiar place or with a familiar mate (or both) are necessary, especially for males who need to compete for territories and a place to nest.

*Nesting site specificity of albatrosses.* No birds forage from more scattered resources and few practice such extreme social monogamy as the wandering albatross. A pair mate "for life" and may return to the same nest year after year for over sixty years. It takes the pair nearly a whole year to produce one chick (78 days of incubation and 250–280 days to feed until fledging). Many pairs nest only once every three or four years. The scattered and widely dispersed food required for these birds constitute the longest (both in distance and time) foraging ranges known. Satellite tracking has revealed that wandering albatrosses in the South Pacific fly on average 1,500 kilometers (km) during a single oceanic foraging trip when feeding their young, and satellite tracking for three consecutive foraging trips of one female who had a chick in the winter revealed trips of 4,173 km (5 days), 3,298 km (4 days), and 6,479 km (8 days).[19] Throughout these apparent wanderings over the open ocean the birds are never lost because after they have secured a load of food, they return directly to the nest in a straight line. Intervals between feedings of the chick by one of the parents may extend to two weeks, and both members of the pair are required to keep the chick fed within the limits of its fasting tolerance. Successive fasting by the chick is alternated with gorging, as a parent may return with up to two kilograms (kg) of food, mainly squid and fish, to regurgitate to the chick. The nearly five pounds of food, for a typically eighteen-pound bird, would be equivalent to a 180-pound person ingesting a meal of fifty pounds (around 200 typical "quarter-pounder" hamburgers).

The albatross parents invest in their chick's future by feeding it enough to endow it with a huge store of fat; it weighs about 160 percent more than one of its parents when it finally leaves the nest. This energy source is a buffer of parental care that tides the young bird over while it learns to become a proficient forager without parents during its life at sea.[20] When in two to four years the parents finally return to their nest to breed, they engage in a long greeting, perhaps courting, ceremony.

Like all other seabirds (and about 92 percent of other bird species), the twenty-four species of albatross are considered to be monogamous because the same individuals return to the same nest site to breed, despite wandering widely in the interim between nestings. Satellite tracking has revealed that albatrosses may cover 184,000 km of travel over the oceans in their first year of life, and then they may wander another seven to fifteen years before breeding. Using still largely unknown mechanisms, they return to the same, often tiny, island where they were born, and after breeding there once, they return in the subsequent breeding attempts to the same nest or nest site they used before.[21] The females select the original nest site, but as in storks, both members of the pair return to the same nest.

The albatrosses' "extreme monogamy" does not mean, however, that they are sexually exclusive. As in other seabirds, a quarter of the young raised by any pair may be sired by a different male than the mate. In case of a mate loss, the males still return to the same nest, but females who lose a mate may re-nest some distance from it but in the same colony.[22] Perhaps this indicates an especially fine-tuned mate choice; perhaps albatrosses recognize deficiencies and choose other partners on the basis of their fitness. We presume that the depth of love is measured by exclusivity, where if one's partner dies he or she will not take another. Although there are reports of widowed albatrosses who did not take a new mate for three years after their mate disappeared (Carl Safina, pers. comm.), we do not know if their reproductive hiatus was due to unavailability of alternate partners, grief, or such factors as age or sickness. "Love" in the sense that we know is measured not just by fidelity to a "house" (nest) or location, but also by a mutual fixation onto each other. Such fixation may not be adaptive outside the context of the narrow locality of the nest itself in a bird like an albatross, which needs to cover vast distances to find food. Two separate pairs of eyes searching in different directions may find more food than the two staying all the time together and thus overlapping their search area.

Most birds' pair bonds are established in spring, and after the nesting season the pairs split up. However, in diverse groups of species the pairs stay together.[23] In western Maine, for example, I routinely see red-breasted nuthatches in pairs in the winter. Pileated woodpeckers also maintain pair bonds through the winter, and at dawn and dusk the paired individuals exchange vocal and drumming signals identical to those they give in the breeding season.[24] Marbled murrelets are commonly observed in pairs at sea throughout the year, and a radio-tracking study by Laura Tranquilla and a team of seven colleagues found that all pairs captured were male-female pairs, and these females were more likely (72 percent) to produce eggs than single females (8 percent).[25] If a compatible partner can readily be found, then there is no compelling or at least obvious adaptive advantage for the pair bond ("love") to persist from one season to the next. However, if mating partners are rare and ready-paired birds get a head start in breeding and cooperate in foraging, then losing a mate is costly, and it is better to stay with the current one.

*Teamwork in ravens.* Some birds may hunt cooperatively to subdue prey, but only ravens stay together year-round as companions and cooperative hunting partners.[26] Common ravens, like storks and albatrosses, also come back year after year to the same nest site and sometimes even renovate the same nest. However, they provide an example of a bird practicing a monogamy that is based and maintained by mutual psychological attachment to each other. As already described in *Brehm's Tierleben* (1893): "The raven belongs to the birds who, once paired, stay always together. If one hears one of the pair, then one only needs to look around to see the other, who will not be far." The raven pair bond is long-lasting. In the Warsaw Zoo, a famous raven couple, Anna and Pavil, built a nest and mated in January 1968, and they continued to nest annually until they had produced seventy-six young by 1987 when someone apparently stole Anna. The zoo then provided Pavil with another female from the Krakow Zoo. He immediately attacked her. Three months later a weakened raven was found walking about in Warsaw, and since she reacted when addressed by the name "Anna," this bird was brought to Pavil. The two greeted each other wildly and with obvious pleasure. They reunited immediately and produced one more clutch of young before Pavil apparently died of old age (he lost feathers and had cataracts). I have observed similar devotion of raven pair members to each other, and similar apparent disdain to a possible replacement of a mate when the first one was missing.

Raven pairs preen each other outside the breeding season and nightly perch near each other. The advantage of them staying together to rear the young in any one cycle is probably not different from that in other songbirds, except being even more relevant because the young take much longer to grow up and require parental guidance for a month or more after they are out of the nest. Second, their ability to stay on territory through the winter eliminates the necessity of fighting for a new site every spring; the home advantage helps to ensure dominance and allows them to start nesting right away when the time is appropriate. Unlike albatrosses, raven pairs hunt and forage together. Raven pairs form coordinated teams in securing prey from hawks and eagles, and eggs from large waterfowl such as swans and herons (Peter Berthold, pers. comm.). One of the pair typically distracts the raptor or waterfowl holding a prey item from the front, while the other comes in from the back and steals the egg or prey.

Recently, ravens working in pairs have started killing and eating lambs in the Alps (Berthold, pers. comm.). The phenomenon began after sheep were no longer raised for wool, ranging freely accompanied by a shepherd and his dog. Instead the sheep were confined in set locations and raised for meat. Ravens could gather among the sedentary sheep bred for meat because they reliably found afterbirths at lambing time. Confining sheep to raise them for meat allowed farmers to keep breeds that could have two lambs rather than the usual one that had to be ready to travel with the flock. In the sheep with two lambs, the first lamb that is born lies still while the mother is birthing the second. At this time, while the mother is preoccupied with birthing the second lamb, the ravens swoop down and kill the just-born defenseless one.

Perhaps even more important, raven pairs act as teams to fend off competitors of their own species. In late winter when ravens nest, animal carcasses are often monopolized by groups of nonbreeding ravens, principally juveniles. All of the meat from a deer carcass, for example, can be removed by them in less than a week, although it would be enough food for a major portion of the nesting season for a pair and their young. A pair of ravens that has found a carcass in their territory actively defends it and hence the cooperation helps to conserve an otherwise ephemeral food bonanza. Furthermore, when one parent is away foraging, the other usually stays to guard the nest, where it sounds the alarm that quickly recruits its mate for help.

*Mate interdependence in hornbills.* The necessity of a pair for a division of labor is obvious in hornbills, where one member of the pair is rendered helpless and totally dependent for its survival on the other. These Old World tropical birds of over fifty species nest in tree holes, and safety of the eggs and young is greatly enhanced, if not ensured, by a curious nesting behavior. Suitable tree holes for nesting are rare, and natural selection has produced the ability not only to modify suboptimal cavities but also to go a step further and almost close off the entrance, making them nearly predator-proof. The female gets sealed in, and her mate then feeds her and their young through a tiny slit until they grow up and until she molts and regrows her flight feathers.

The female ensures, prior to her entombment, that her mate will be a reliable provider. She solicits food from him by quivering wings, fluffed head feathers, and begging vocalizations. If she gets the right foods and sufficient amounts of it to guarantee his devotion, she selects a tree cavity as a nest site and spends increasing amounts of time in it as he continues to feed her. Increasingly she stays inside, and the male then ingests soil, mud, and sticky foodstuffs, which he brings and regurgitates to her. She applies this cementing material to the nest entrance, where it forms a hard wall. Eventually only a slit is left, through which the male then passes food to her. If he were to leave her or something should disable him early in the nesting attempt, she would still be able to peck through the nest entrance wall and fly away. But eventually her nesting will require her to shed her flight feathers, and she would then be helpless for months. If the male continues to feed her, and if nothing disturbs the nest hole for about a week or more before she starts to lay her eggs, then it is likely a safe place, and she begins to lay. Within a day or two of laying the first egg she becomes totally committed, because she molts her tail and wing feathers. She will then be flightless for months and will have to stay in the nest hole.

The eggs are laid three to five days apart, and with up to eight in a clutch, it takes over a month to lay them. It requires another month to incubate them and at least as long until the young are flight-ready. By then she will have grown a new set of flight feathers. After she and all her young are flight-ready and can leave their prison, it takes still a few more days of hard pecking before mother and chicks escape the nest cavity.

*Multitasking in sandgrouse.* Inaccessibility of the eggs and young from predators can be achieved by various means, and the African ground-nesting sandgrouse find relative safety for their young by breeding deep in inhospitable deserts. Sandgrouse are not grouse, but belong to their own family, Pteroclididae. Their choice of nesting grounds ensures that predators are few because of the heat and absence of nearby water. As for the hornbill, it is the activity of the male that makes such extreme nesting possible; he serves as a water carrier while his mate must stay with the chicks and shade them from the sun.

The sandgrouses' clutch of three eggs is laid on the open ground in a shallow depression, and the pair takes turns incubating. The young are precocial but cannot fly at all until four or five weeks, and not well until they are about two months old. All of this time they are dependent on their parents not only for food but also for shade and water for evaporative cooling to survive the searing heat. Their only source of water may be up to eighty kilometers distant, and although these birds are fast flyers and can imbibe water very quickly once they reach it (five to ten seconds, by sucking), even minutes away from the young in the hot sun puts them at risk of dying from overheating. The solution to this problem is that the male has evolved specialized feathers on his belly with fine filaments that have a water-holding mechanism and act like a sponge.[27] These allow him to soak up about forty milliliters of water by wading out into standing water and rocking his body back and forth. Even after a prolonged flight of thirty kilometers he brings fifteen to twenty milliliters of water back to his young, who crowd under him and then strip the water from his belly feathers with their beaks.[28] He makes a daily trip for at least two months until they themselves can fly to the waterhole to drink.

*Flamingos fill a niche.* Both lesser and greater flamingos also breed out of reach of predators by inhabiting some of the hottest places on Earth—the middle of mudflats that, at Lake Natron in Tanzania, stretch over 1,000 km². Temperatures there may reach 65 °C. As many as 3 million pairs may nest in the same colony, where each pair builds a mud nest that reaches about a foot above the salt-saturated water and mud. The birds feed near the nests by filtering microorganisms such as common blue-green algae out of the salt solution. Few other animals can drink the salty water, but the flamingos can do so because they have special salt-excreting glands. The young are fed like young pigeons are, by a secretion from the crop. When feeding on red bacterial halophiles in Lake Natron in Tanzania, this crop "milk" is crimson and resembles blood.

*Egg cooling by Egyptian plovers.* The Egyptian plover is not a plover nor does this African bird still occur in Egypt. It belongs to the coursers, Cursoriinae, a group of ground-living birds that superficially resemble shorebirds. Parenting in these "plovers," as determined by the late UCLA ornithologist Thomas Howell in a detailed study of these birds in Ethiopia, requires multiple tasks that involve male/female cooperation. Unlike sandgrouse, these birds lay their clutch of two to three eggs in a scrape of loose sand by a riverbank, and gain protection from both predators and heat by burying their eggs and sometimes their young with that sand. Predators, principally crows and kites, are common, and so the combination of heat and predators requires vigilance in two directions at once. Apparently not only one nest scrape suffices, because a pair may make dozens of scrapes before finally settling on a nest site.[29] I speculate that this behavior is not mere indecision. It probably acts to confuse or discourage the crows; a crow might have to dig in the sand in many places in order to find the one spot where the eggs are actually hidden. It might not persist in digging in the sand unless it is rewarded.

On cool days and at night the parents incubate the eggs, but as temperatures rise along with the sun in the morning, they cover them with sand to varying depths and let the sand continue the "incubation." Soon, however, shading is not adequate protection from the heat, as air temperatures may reach 45 °C in the shade and 50 °C in the sun. Both adults then take frequent trips to the river where they wade out into shallow water, crouch down to partially immerse, and then rock back and forth from front to back until they have soaked up water. Then they return to the nest and crouch down to let the water moisten the sand so that evaporation keeps the eggs from overheating. Throughout their egg care the birds monitor the nest temperature (apparently with their beaks). The young quickly leave the nest after they hatch, but they still require direct care from both parents, who remain alert for predators and sound an alarm for the young to crouch down when one is seen approaching. While one parent distracts the predator, the other quickly covers them with sand. The Kentish plover in southern Turkey adds nest material that camouflages and shields the eggs from the sun in the day, then uncovers them at night to incubate.[30]

*Parenting in thermometer birds.* Sandgrouse, sandpipers, geese, grouse, and penguins may lavish parental attention on their cute, fuzzy chicks, but the twenty-two species of Megapodidae devote just as much time and attention to their eggs, although at first glance their devotion may not be apparent.[31] These birds bury their eggs and never see their young. Megapode young are "super-precocial," and after a struggle lasting from two to fifteen hours, they hatch into a compost heap and dig themselves out unaided. In a few moments they are ready to run, and having ready-formed flight feathers, they can also fly within a day of hatching. They maintain an independent solitary life, having no contact with either parents or siblings. Nevertheless, the huge effort to produce them requires both parents.

Since parental care in the megapodes is lavished almost entirely on eggs that are never incubated, it might appear that these chickenlike birds from the Australo-Pacific region have opted for a lazy parenting style. It might have started out this way, possibly when females covered their eggs (as many birds do when they leave the nest). But it didn't end that way. In one species, the malleefowl of Australia, the male builds a massive mound about one meter high and five meters in diameter. (Some other species nest in warm, sun-heated sand, and one lays its eggs in tunnels dug toward hot volcanic streams.)

Megapodes live in warm regions, and presumably during their early evolution if a female had her nest in a warm place, she could have stayed away longer. If she was then able to find more food, she could also lay more eggs. After the making and manipulation of a compost heap evolved, a male who helped out in the egg care allowed his mate more time to feed, and she could again lay more eggs—eggs that he fertilized.

We don't know the precise intervening steps, but today the male malleefowl digs a pit about a meter deep. He then gathers leaves and other organic debris from the surroundings and scrapes it into the pit. After it is moistened by rain, he covers it with sand and soil. Microbial action takes over to produce heat, and he at intervals opens the mound and takes its temperature with his thermally sensitive bill and tongue. When the temperature in the mound reaches 33 °C the hen starts laying eggs. Now the partnership between the pair continues, as he daily monitors the mound by opening it and probing around the egg or eggs, rearranging the mound material to maintain it at a constant temperature. Meanwhile, the hen is free to forage rather than incubate, and she can therefore continue to lay her large eggs (each about 10 percent of body weight) for about half a year, usually laying one egg per week. All that time the male is homebound, busily involved in keeping his compost heap at the right temperature.

*Parenting by two or by one.* Few sights inspire our empathy for unrivaled devotion more than a married pair of geese or swans accompanied by their young. Although the young are precocial and feed themselves, from the time that they hatch from the eggs they are highly vulnerable to predators, and the adults are large enough to serve as credible defenders. The young are never out of sight of their parents or alloparents while they are growing up, and even beyond. In the fall their parents lead them on the traditional migration route. A mother duck and her ducklings push our cuteness button, but the male who leaves this lovely family is thought less admirable. However, in contrast to geese and swans, male ducks cannot credibly defend their family. Probably as a consequence, ducks serve their reproductive goals better by not drawing attention to themselves when associating with their young. An accompanying male duck might instead serve as bait to draw a predator. He must therefore stay away from the family. He probably cannot improve his reproductive effort by helping them, and by having the opportunity to leave, he can try to mate with other females. Polygyny is thus made possible. Geese and ducks, though otherwise closely related, have evolved highly contrasting parenting styles, and thus they provide a window into the evolution of mating systems.

Having a partner active in the defense of nest and young does not always require large body size, however. The nest defense of smaller birds necessarily involves still other, sometimes active, strategies; they can use themselves as bait to lure a predator away from their offspring. They feign injury, making themselves seem even more vulnerable than they really are. This strategy is used almost universally from tiny ground-nesting songbirds to grouse, shorebirds, doves, and female ducks. Males are generally less involved, although some shorebird males can also make themselves useful by serving as sentinels to warn their mates of danger. Then she can lead the young away or warn them so that they hide. In the semipalmated plover, for example, the male gives a distant warning when he sees a predatory bird, upon which the young instantly hide (M. McVeigh, pers. comm.).

*Parenting by two sitting tight.* Like the monogamous partnerships of the hornbills and the geese and swans, those of doves may be driven by predation. But as I observed in a pair of mourning doves nesting on an apple tree near my window, it is done in an entirely different way. The "genius" of a dove's nest is that it does not look like a nest at all. It's just a few twigs and debris thrown together into a crude platform, and when the bird perches on it, it is not obvious that there is a nest there, much less that she might be hiding two pearly white eggs or two young. From the day that the first egg was laid until the young fledged, I never once saw either the eggs or the two young without a parent covering them. The weather was warm and sunny during most of the fourteen days of incubation and the twelve days that it took for the hatchlings to be able to fly, so there was no need for a parent to keep the fully feathered young warm, yet there was never a moment during the day that first the eggs and then the chicks were not covered. The bird's tenacity was remarkable. I was able to climb the tree and photograph him or her (in this species both members of the pair incubate) on numerous occasions without flushing the bird from the nest. After the eggs hatched, the attending bird continued to sit on the young and never budged once as far as I could tell. (I checked numerous times every day.) The feces accumulated at the nest as well, again reinforcing the strategy of reducing all movement and activity that might reveal that the perched bird was actually associated with a nest. Had the brooding bird flushed, and had I been a predator, the pearly white eggs or the helpless young would have been an instant target. However, many birds in the tropics who build respectable nests also have only two eggs per clutch, and one hypothesis is that the low egg number is an adaptation against predators because it reduces the frequency of nest visits to feed the young. It also reduces the chances of the parents leading a predator to the nest.

In most perching birds both parents are required to find sufficient food to support the heavy demands of the growing young, on whom there is a strong premium to grow, passing quickly through the very vulnerable and helpless juvenile period. Having only two young per clutch may be an adaptation for minimizing the frequency of nest visits, and it thus helps their survival by keeping the nest location secret. Doves, who feed their young with masses of food transferred from their full crop to fill their youngs', may restrict nest visits to just one per day. Furthermore, permitting one of the adults to be permanently available to shield the nest and its contents would be important.

I found the tenacity with which the mourning dove sat on the eggs and young and refused to budge astounding. Doves invariably fly up when I come within about fifteen meters of them, but the bird on the nest did not flush even as I climbed four and a half meters up into the apple tree where the nest was, and where I then approached along a long, semi-horizontal branch to take repeated photographs holding the camera lens within twenty centimeters of the sitting bird. I never saw it make a motion, and it never blinked an eye nor shifted the position of its head even slightly. It only flushed off the young on the nest after they could fly well (uphill). I then noticed that the adult and the two young had been sitting on a huge accumulation of feces, so the attending bird had been hiding that as well. My observations of these birds—their minimal nest; covering of eggs, young, and feces; freezing of all motion; absence of any begging vocalizations of the chicks; very fast growth; and staying in the nest until expert flight was possible—all point to predator avoidance as a unifying strategy. Predator avoidance may also explain the structure of the nest, the number of eggs, and the crop feeding, as well as the obsessive nest sitting. This strategy requires the presence of the male to provision the sitting female and to relieve her at the nest. Without a partner, it would not be possible.

Hummingbirds also lay only two white eggs, yet there is no partnership between the male and the female. The male leaves after mating, and all the nesting is accomplished by the female alone. I think the key to the difference between this system and the doves' is the near-invisibility of the hummingbird's nest—a tiny cup that is camouflaged with flecks of lichens and resembles a knot on a limb. The deep nest cup is lined with white plant down that hides the two pea-sized white eggs. Helping to make nest detection difficult may reduce predator pressure and help make the male's presence less important (though still useful if he could help incubate) because the female can take brief respites from the nest to feed herself.

*Teamwork in distraction.* Birds that are parasitized by cuckoos often attack the female before or while she is attempting to insert her egg. In some cuckoos, such as the Jacobin cuckoo, pair partnerships are needed to bring off the trick of egg insertion.[32] The male cuckoo (who looks to us identical to the female) is the first to approach the target host nest. When the nest owners then start to attack him and are preoccupied with him, he leads them away, and in that moment the female rushes to the nest and dumps her egg.

The cuckoo pair may have to stay together for a long time, since they typically lay large clutches and may parasitize more than a dozen nests (not all their eggs will be accepted). Indeed, in the spotted cuckoo pairs accompany each other constantly.[33] They show all signs of being bonded by their behavior of perching side by side, touching bills and vocalizing to each other in soft tones.

Copulation in this species is a much more elaborate ritual than in most birds, in which it is a perfunctory affair lasting about two seconds. In most other species, the male hops on, scoots backward to make cloacal contact, and then hops off. The male spotted cuckoo, however, finds an insect before mounting his mate, and during the copulation he holds it down to her. She grasps the insect while he still holds it as he balances on her back. He relinquishes it to her only after a lengthy copulation of about two minutes. These behaviors suggest that the copulation serves as a bonding ceremony that helps maintain the monogamous pair bond. The greater roadrunner, a monogamous species of cuckoo who mates for life, also copulates more than is necessary for fertilization. Roadrunners copulate at the site where they will build their nest, and they continue to do so even after their clutch has been completed, hence suggesting their activities function to bond the pair.[34]

*To hold in order to have.* Just as a suitable nest site could be rare and become a limiting resource, so could a mate. Having a mate already on hand saves a lot of time and effort because it reduces the potentially lengthy task of acquiring one every reproductive cycle. Most migratory birds return to their breeding grounds early and try to get a head start. Red-winged blackbird males, for example, who need to acquire territories to attract mates, arrive three or more weeks before the females show up. Herring gull first-time breeders form pairs at the breeding grounds while former breeders show up with their mates. Still others, like the common raven, stay paired all year round. But hanging on to a mate is especially critical if the species is dispersed and rare. Here in Vermont there is a pair of sandhill cranes that has been breeding for four years (at least since 2004) at a site within a half-hour drive from my home. This pair is almost a thousand miles out of the normal range of the species. As soon as one of the mates is without the other, there would be slim chances of either continuing to breed. It is therefore advantageous for the two to remain attached to each other. In other words, the behavioral bond of these birds keeps them available for each other, possibly for decades, thus permitting them to produce numerous offspring. Bonding to the rare mate is important, especially when the chances of finding a new mate at each breeding cycle are remote.

*"Ordinary" social monogamy.* The examples given in this chapter highlight some of the many possible situations where monogamy is based on the special services that a male can or must provide in order to

enhance his reproductive potential. In the majority of birds with help-less altricial young needing parental care, the essential role of the male may be more pedestrian, though just as crucial. A male who in-seminates a female gains no reproductive benefit unless the resulting young from the mating grow up and in turn reproduce. Parenting by most birds requires a huge parental investment in the female to pro-duce and incubate the eggs, and then to feed fast-growing altricial young. Food provisioning is a limiting factor in fledgling production and survival in perhaps most birds, and the male's investment in that provisioning is essential for the pair's, and ultimately, his reproduc-tion. In many cases the male's efforts equal and possibly exceed that of the female. Thus it is worth the female's time and attention to se-cure the right male, and courting provides a filter that narrows her choices. Courting behavior is at least as much if not more important to the female, however, when the male provides no material benefits.

## 3

POLYGYNY

AND

POLYANDRY

*AUGUST 14, 2005.* FOUR DAYS AGO, ONE

after another of the baby house wrens hopped to the entrance

of the bird house, peeked out, and then flew off. This was the

second brood of the year from that box. Afterward I heard the

noisy wren family almost continually in the bushes all around

our house. But yesterday I heard baby wrens chirping from an-

other bird box near the first—the one from which tree swallows

had fledged their brood in July. I looked into the box and found it, in typical house-wren fashion, stuffed nearly to the top with sticks and lined with a pile of brown feathers on which hunkered a clutch of pinfeathered baby wrens. This (third) clutch had been carried along with the second, even though only one male had sung here. Apparently he was a polygynist—he had mated with a second female and with her had a third clutch overlapping with his first female's brood. If so, he was a mild one—not like the rooster in our yard who mates with every hen; each hen incubating and tending her many chicks alone.

Some environments and conditions allow or encourage males to forgo parenting and provide only genes to females so that polygyny is practiced as an option or can evolve to be a primary reproductive strategy. Other situations demand so much parenting that one male isn't enough, which encourages females to become polyandrous. The asymmetries between the opportunities and the amounts and kinds of what is demanded of males and females for reproduction are large and varied, and they give us examples of "problem solving" by evolution. In many cases we can deduce the problem that has been solved by examining the solution, but sometimes the solution seems so convoluted (and interesting) that it is hard to ascertain what the original problem might have been.

Presumably polygyny could evolve either where the females have so few and undemanding young that they can feed them by themselves, so that males become superfluous as help, or where the young are precocial and do not need to be fed or cared for by two parents. Males, with "nothing to do," are then free to expend considerable energy in trying to secure additional mates, which can result in intense sexual selection for displays involving bright feathers and complicated song and dance routines, most famously those of peacocks, some grouse, widowfinches, manakins, and birds of paradise. The females inspect and compare the males, choose from among them, then crouch in front of them and are mated. One male can mate with many females, and if the criteria that make an attractive mate are heritable, then it is of evolutionary advantage for the female to preferentially mate with those males who get most of the matings because it will translate, through her more attractive sons into having many descendants—what amounts to "runaway" selection.

The best way for a female to evaluate the most attractive males by the criteria that other females use is to see them in a lineup. Males are therefore "forced" to compete against each other in direct face-offs, because if they don't strut their stuff to a female audience they get no matings. In such male assemblies ("leks"), a very few males then get all the matings and the rest get few or none. In such polygyny, as it is practiced by various species of birds of paradise and Phasanidae, the females typically choose the most spectacular males of a specific type. In the greater sage grouse of western North America, for example, the superior display that all eligible males must have includes two giant bare yellow inflatable protuberances that pop and jiggle out of their white breast plumage to the accompaniment of a loud bubbling, popping sound. Since the females flock to attend such leks to compare and evaluate these displays, lone males and those who don't conform are ignored, so the males are forced to compete head-on.

Ironically but perhaps not surprisingly due to the intensity of the competition, in some species males form cooperative partnerships. One coalition of males competes against another, much like one pool of male wood frogs competes against another in their combined calling volume. In the long-tailed manakins of Central America, the male-male partnerships for competition involve at least two males, who have taken four years to acquire their adult plumage. These beautifully blue- and crimson-attired males whistle in almost perfect synchrony, and in their cooperative display the couple alternates their movements; as one lands on their horizontal display perch, the other jumps and performs a hovering "butterfly" flight. Females that are attracted to the pair's performance nevertheless compare the participants. Both are adults, but one is of a higher status than the other. When the alpha bird of the pair finds a female is impressed and ready to mate, he makes a call that induces the beta male to leave, and the alpha is then left to mate. There is no reciprocity, nor are the two siblings; the dominant male does virtually all the mating throughout the whole breeding season. In one study the alpha male performed 259 of the 263 matings observed.[1] There is therefore very little or nothing in it for the beta male unless he has patience and longevity; he may get almost no matings in any one year, but it pays him to stay and help because he stands to inherit the lek after the reign of the alpha is over.

Polygyny has taken a bizarre twist in the ruff, a species of Arctic sandpiper. The males, as in numerous bird and other animal species, also display in leks. As in the manakin, two males may cooperate, and females are preferentially attracted to leks with the pair. However, in the ruffs' system there is not just one ideal type that all males must embody in order to mate. Instead, the males at any one lek are arrayed in two conspicuously different and showy feather-dress types (and there is individual variation of this dress as well). One general type of male has a large ruff of white feathers around the neck, and another has a dark (black or brown) ruff.

In most lekking birds where the males are polygynous, the females are "monogamous"—they mate with only one male, and as already mentioned, they are arguably more choosy than typically monogamous birds since most of them mate with the same male. But in ruffs, the females, although they are also highly choosy, are nevertheless polygamous—each mates with several males (apparently only to diversify their genetic contribution). After mating they leave the males and incubate their eggs alone. They also take care of the young without the help of a male for the month to six weeks during which they must ensure they are fed.

The two types of ruff males differ not just in feather gown but also in behavior. The dark males are territorial, and the light ones are "satellites" of the dark who move freely from the lek of one dark male to another's. Any one lek measures about one square meter, and it is defended by the territorial dark-ruffed male against other males that look like him. However, since the females find leks that contain both a dark and a light-colored bird more attractive than one that has only one dark bird, these dark males are required to actively recruit the nonterritorial light-colored satellite males to form their temporary alliance. However, after the satellite male "helper" has been recruited, the two then become competitors; either one may mate with any of the attracted females. Thus, both morphs continue to coexist in the population.

The existence of two male morphs in the ruff had fascinated biologists for over a half-century, but recently Joop Jukema and Theunis Piersma made the very surprising discovery of a third male ruff morph.[2] These males had never been noticed before because they closely mimic the nonshowy females in both garb and general behavior. Even other male ruffs are apparently fooled by them and find them convincing female mimics: they sometimes mount them. There is, however, no doubt that they are in fact males, because the dissections by Jukema and Piersma revealed that they have testes. Big ones. Indeed, during the breeding season, the testes of these femalelike males were two to three times larger than those of the showy males. These large cojones suggest that they may get fewer matings, but they make up for it by sperm competition.

The female-mimicking "sneaker" morph apparently exploits the other two, not by display but by insinuating himself on the sly next to the displaying males and mating with the primed females attracted to the others. And also unlike the behavior of most other lekking species, all these males are silent. Their mating strategy of penetrating dominant males' defenses is found in other animals, especially fish, where there is also intense competition among males for matings and opportunity for sneakers. But in all other species this strategy is optional: the animal changes appearance and behavior depending on conditions. In the ruff, on the other hand, the sneaker morph, like the other two morphs, is permanent; the animals are born that way.

How and why the ruff system has evolved has long been a puzzle. However, it may not be as mysterious as it may seem. I speculate that the black ruff was the original male, and as in other lekking systems it evolved to be conspicuous. And as with any lekking system in which there are many males without a mate, there was room for a competitive "sneaker" strategy for some males to breach the breeding circle by ducking under the radar in the presence of many females. The females are already primed to mate, and the "sneakers" take advantage of the situation without being noticed by the harem master. But there is also another way. Suppose a ruffed dominant male has a mutation that gives him a white ruff. The dominant, conventionally attired male may then not regard him as a competitor, and he is thus admitted with little challenge on the basis of being considered inconsequential. However, to the females who are attracted to a conspicuous mating display, there is perhaps an added inducement of more total showy display at the lek. Experimentally, it has been shown in zebra finches that the addition of an arbitrary showy ornament makes the male finches even more attractive than normal males. And so it is in another species as well—the white-ruffed males become ornaments or helpers that a dark-ruffed ruff (or his lek) needs to recruit in order to compete with a neighboring lek, where the added inducement may be missing.

In the polygynous lek systems there is no conflict among the females, because sperm is plentiful enough. But when the males offer more costly resources, such as food, then female-female conflict is almost inevitable, as seen in polygynous reed warblers. As in most songbirds, female reed warblers choose a male in part on the basis of the territory he holds, which is an indicator of how well he will be able to provide for her kids. Not surprisingly, when resources are strongly clumped, such as in a marsh, one male can dominate a rich local area, and in the great reed warbler of Europe several females may settle in his territory. He allocates most of his help to the first female that arrives.

Three Swedish biologists, Bengt Hansson, Staffan Bensch, and Dennis Hasselquist from Lund University found that the male reed warbler's primary female suffered three times more predation on her eggs than the secondary females.[3] To find out why, they wanted to identify the presumed predators, so they set up fake nests containing fake eggs whose surface easily scratched to leave tooth or bill marks that could be matched to presumptive predators. They found out that the predators were not cuckoos or mice or crows, as could easily have been assumed. Instead, the eggs had been destroyed by other greater reed warblers. They deduced that the secondary females were destroying their competitor females' nests to thus gain the help of the male, who would then feed their own offspring instead. Such, then, is the cost of polygyny to some females, although it is advantageous to her male (but of course not to other males that most probably end up with no mate).

Ostriches also practice a type of polygyny, although in a very different way than most polygynous birds. They live in small groups consisting of a territorial male with two to seven hens. He prepares a nest scrape and all his females lay eggs in it. Up to sixty eggs can end up in one nest. The dominant female incubates, and although it would be to her advantage to hatch as many eggs of the other birds as possible (because the others' chicks would dilute the chances of a predator-caught chick being one of her own), when there are more eggs than she can incubate she recognizes her own eggs and preferentially rolls them off to the side. The female incubates during the day, and the more brightly colored and black male incubates at night.

Partners can be competitors at the same time. Female Eurasian penduline tits, for example, choose a male who provides her a large, beautifully made nest, but then after she starts to lay her eggs in it he deserts her to build another nest to attract another female, repeating the process. But she has her own strategy to circumvent his and try to keep him around, thus helping her enhance her reproductive potential. She may lay just two of her typically six eggs of a clutch and then cover them up with nest lining. He stays to mate with her, presumably to fertilize the whole clutch, but when she has laid the whole clutch she may leave. He would lose all of his investment if he left, so he ends up stuck with the eggs while she finds another male advertising another nest, and so she gets off two clutches. Thus while he is polygynous, she is polyandrous, and so they bring off more than one clutch per year.

Perhaps no birds face greater time constraints to raise their young than Arctic shorebirds, but during the short Arctic summer they have a temporary abundance of food, and they can feed for twenty-four hours every day.[4] An Arctic summer may not extend much beyond a month, and during that time the birds have to find and court mates, build their nests, lay and incubate eggs, and rear the young. The young must be ready not just to fly but to migrate thousands of miles. Remarkably, in shorebirds not just one but several contrasting parenting styles have evolved to take advantage of the opportunity and also solve the problems associated with it.

Most shorebirds save time by being ready to nest shortly after arriving from the southern continents where they spent the winter. Many (such as the black-tailed godwit and the green sandpiper) come already paired to their northern breeding grounds. With one or two exceptions the shorebirds are all ground-nesters, who don't need much time to build a nest—either a mere scrape suffices, or males may dig several nest molds, and the female then chooses one. She adds a little grass, and he may then leave before she lays her four eggs. These eggs are large relative to the bird's body size, so the young have a head start even before they leave the nest, which is almost immediately after they hatch, as soon as their dense down is dry. In most species (*Tringa* spp. sandpipers, turnstones, lapwings, sanderlings, little stints, and godwits) both members of the pair take turns incubating, and then both also accompany and protect the young. In others (woodcocks and common snipes) only the female incubates, or the male may or may not incubate (curlew sandpipers), or may do so only occasionally (golden plovers).

The female sandpipers' role in the care of the young after they hatch varies as well, and their behavior has led, in a number of species, to interesting innovations in polyandry and cooperative parenting. In turnstones, for example, both parents incubate, but the female abandons the young shortly after they hatch. The male takes over and leads them around. In Temminick's stints, this pattern has taken on an additional twist—the female lays two clutches of eggs in rapid succession. Unlike a turnstone, she leaves the first clutch after laying it, so her mate is then stuck with incubating as well as leading the young after they hatch. She, meanwhile, lays a second clutch soon after leaving the first, and since her mate is then busy with her first clutch, she must incubate this one to completion. Then she also leads the young. Sometimes, however, she finds a second male for her second clutch, and that second clutch may then be attended by either one or both parents. In North America the spotted sandpiper, though usually monogamous, may also be sequentially polyandrous. That is, the female may take successive mates as she leaves each male to attend to the next clutch. In this arrangement, males become a limiting resource for females, and contrary to most birds, the females return north to the breeding grounds before the males. They then compete with other females for territories and defend them from each other.[5] When the males arrive they are attracted to the females' territories, and the females then mate with them and lay their typical shorebird clutch of four eggs, bringing off two or more clutches of young in one season whereas otherwise they could only have one.

A male spotted sandpiper's evolved willingness to incubate eggs should be contingent on his assurance of paternity. However, male paternity is never assured, and females may not need to be choosy in a partner, so long as he will incubate her eggs. A female appears to placate a potential mate by copulating with him frequently, which would increase his chances of paternity and hence make him willing to incubate the eggs and then lead the young around. However, she nevertheless exerts mate choice beyond just recruiting a willing egg and baby sitter. She may practice cuckoldry by storing sperm from an earlier, preferred male. She mates with the first male to fertilize her eggs, and then allows the second one to incubate them.[6] The male sandpiper that refuses to incubate the eggs that his mating partner abandons would have no descendants, so he has little choice.

Phalaropes practice a type of polyandry similar to that of spotted sandpipers. It differs in that the females court the males directly rather than attracting them to defended territories.[7] Once a female has a mate, she lays her eggs in the nest he builds, and then she leaves him and finds another mate to whom she donates the second clutch of eggs. The polyandrous parenting style in phalaropes is likely a consequence of a large but brief local food supply. Shorebirds seem to be constrained to laying and incubating clutches of four large eggs. The female gets off eight or twelve offspring because the male can manage the young. Males can manage this only because the young are precocial and don't require two parents to feed them. As in the spotted sandpipers, since the reproductive success of the females depends on finding a male who is willing to take care of her eggs and young, there is strong selective pressure to compete with other females for the parenting services of males, which may explain why the phalarope females are larger and more showy in plumage and vocalizations than males, and why (unlike other shorebirds), the females play the leading role in courting, nest defense, and choosing a nest site.

Aside from shorebirds' polyandry, a curious parallel system has been revealed in a cuckoo from tropical Africa, the black coucal. In most coucals both parents share the parenting duties. However, the one to seven eggs of the black coucal are incubated exclusively by the male in his own nest, and he also is the sole provider for the altricial young.[8] Males have only one mate, but females have several, each with their own nest. As with the shorebirds, there is "sex-role reversal"—the females are 69 percent heavier than the males, and they are noisy and defend the territory. Males have higher levels of circulating testosterone than females, both during the breeding season and in the nestling stage, and the sex-role reversal is apparently not due to this species' anomaly of a missing left testis.[9] Reasons for the unique parenting system (and lack of a testis) in these coucals are obscure. Unlike the sandpipers', the coucal's breeding season is long, and there are not likely serious time constraints. It is speculated instead that high nest predation may have favored females who could produce a number of clutches in quick succession. However, local food abundance would also have been necessary in order for the males to be the sole food providers for the young.[10]

Food amount and distribution has long been thought to affect mating systems.[11] It continues to generate much interest, and the effects of food distribution on the mating system as such can probably be seen most clearly within a single species where monogamy, polygyny, and polyandry are all involved in response to changing food supply. One recent study on this topic was conducted within sight of my home in Vermont (on Mount Mansfield) on an enigmatic, little-known bird, the Bicknell's thrush. It was not officially recognized as a new species until 1995.[12]

The Bicknell's thrush breeds in dense spruce-fir thickets on isolated mountaintops in northeastern North America. It lives in an environment where it encounters frequent strong winds, near-freezing temperatures, heavy rain, and marginal food supply (insects). A female of this species typically lays only one set of four eggs per summer, and if she is lucky, brings off the one clutch. To accomplish even that, it turns out, involves a remarkable breeding strategy in which monogamy involving the commitment of one male, such as that practiced by most thrushes and other perching birds, is usually inadequate. Female Bicknell's thrushes usually have more than one mate; each nest has only one female, but the young in it are sired by several males and several males also help feed the young.

The main part of this story was unraveled in a tour de force of work (and fun?) by James E. Getz from the State University of New York and Kent P. McFarlane and Christopher C. Rimmer from the Vermont Institute of Natural Science, with the aid of a small army of twenty eager and able assistants who helped in finding and monitoring nests and catching and marking adult birds with individually identified color-coded rings.[13] They then worked in the laboratory with molecular techniques to determine relatedness and parentage of broods. They found out that in these thrushes, although they superficially appeared to be monogamous pairs, the females were often polyandrous and the males polygynous. In their study of eighteen broods, only four consisted of the traditional male-female pairs. The other fourteen broods were each attended by one female, but with the assistance of two to four males who had also mated with them (as determined by molecular techniques to evaluate parentage of the young). Thirteen males also fed the broods in which they had sired young.

Optimization theory predicts that males should prefer monogamy over having polyandrous females so that they could be assured of the paternity of all of the young that they help feed. But assurance of paternity would add a considerable cost—mate-guarding—and it may be impossible for the Bicknell's thrush in a foggy environment with dense thickets. Much attention is required to secure scarce food, leaving little time for other activities. Where mate-guarding is not possible but moving around is instead required, the males then mate with several females and offer help taking care of the young of their mates. The females, in turn, "should" mate with several males to thus coerce them to help raise her (their) young. That is, by being polyandrous females gain more support in raising the kids, and the males, by being polygynous, make up for what they lose by relaxing their mate-guarding.

A follow-up study by Allen M. Strong from the University of Vermont with Chris Rimmer and Kent McFarlane determined that the Bicknell's unique mating system is indeed a response to food re-source distribution.[14] Females hold small exclusive territories and males have large overlapping home ranges. By having several males, the female effectively increases the area from which food is collected by securing food outside her home range. Her mating decisions are thought to be influenced by the profitability of her home range prior to egg laying. Apparently she solicits copulations from other males when food is scarce, and then these males continue to return to her and her nest and help with the child-rearing. The more food she has, the more she contracts her territory, and the more likely she is to be monogamous (as we also may see in Nick Davies' studies of dunnocks). In contrast, in the spotted sandpipers' polyandrous system a female has multiple *sequential* males, and she can do it because the young are precocial and can feed themselves, given a brief period of ample food. In the thrush, on the other hand, there are several concurrent males because there is food scarcity and the altricial young have to be provided with much food.

A recent study by Dustin R. Rubenstein of the superb starling in eastern Africa also highlights the flexibility and adaptability of bird mating systems to environmental conditions.[15] The superb starling lives in arid regions where rains are unpredictable and food is often scarce. They have adapted to such conditions, as have numerous other birds, by living in cooperative groups of several pairs with numerous helpers at the nest who in hard times share the burden of finding enough food for the young. A single pair could not manage this alone. Pairs are mostly strictly monogamous over their lives. However, some females "cheat," seeking and finding copulations outside the pair bond despite their mates' guarding behavior. These females may be classified as "promiscuous" because some mate with males who are in the pool of nonbreeders of their social group. This mating is apparently strategic because these males become helpers at their nests. Such matings occur ten times more frequently when the group is in poor habitat where helpers at the nest are needed, and they are very rare when helpers are not needed. As a species, the starling may thus be described as polyandrous because the females may mate with several males. It could also be described as monogamous or promiscuous. It all depends on what one chooses to focus on—is the cup half empty, or half full?

## 4

### PENGUINS AND US

AN EMPEROR PENGUIN IS, AT FOUR FEET

tall and weighing up to 40 kg (88 pounds), the largest penguin. It

may dive to depths of 550 m (1,800 ft), and it walks upright in

bipedal gait with its stubby wings swinging side to side like a

strolling human's arms. It wears a gaudy costume that looks

more like a skin-tight swimsuit than a layer of feathers. And it

may walk at a seemingly leisurely but steady pace, usually in a

long line of others, for a hundred kilometers or more to and from its nesting colonies on the winter ice of the Antarctic Ocean. It does not build a nest, and it is the only bird in the world that never ever touches land. It nests on the ice in the extreme South in the depths of winter where it cares for its single egg, and then chick, in continuous darkness, while encountering average temperatures of $-20$ °C (and enduring sometimes $-50$ °C in hurricane-force winds). It is popularly known from *March of the Penguins,* the blockbuster movie hit of 2005 narrated by Morgan Freeman and touted as "a heartwarming story about family and the power of love."

In this movie, the filmmaker Luc Jacquet faced down the demon, if not the taboo, of anthropomorphizing an animal. And I, as a biologist, was asked by the *New York Times* to write an editorial about *March of the Penguins* (August 26, 2005). Naturally in my essay I had to comment on Mr. Freeman's use of the word "love" in the context of the penguin's behavior. I wrote:

> The unspoken rule is that this four-letter word is to be applied only to one creature on earth, *Homo sapiens.* But why? A look at the larger picture shows this presumption of exclusivity is utterly unproved. In a broad physiological sense, we are practically identical not only with other mammals but also with birds—muscle for muscle, eye for eye, nerve for nerve, lung for lung, brain for brain, hormone for hormone—except for difference of detail of particular design specifications. Functionally, I suspect love is an often temporary chemical imbalance [relative to when not "in love"] of the brain induced by sensory stimuli that causes us to maintain focus on something that carries an adaptive agenda. Love is an adaptive feeling or emotion—like hate, jealousy, hunger, thirst—necessary where rationality alone would not suffice to carry the day. Could rationality alone induce a penguin to trek 70 miles over the ice in order to

mate and then balance an egg on his toes while fasting for four months in total darkness and enduring temperatures of minus 80 degrees Fahrenheit and gusts of wind of up to 100 miles an hour? And bear in mind that this 5-year-old [minimum age at first breeding] penguin has just returned to the place of its birth from the open sea, and thus has never seen an egg in its life, and could not possibly have any idea what it is or why it must be kept warm. Any rational penguin would eventually say, "To heck with this thing, I'm going back for a swim and eat my fill of fish."

I saw my editorial more as a basic undergraduate lesson in biology than an opinion piece, but the response that it generated was astounding in its volume, vehemence, and variety. I had expected minor criticism from a few professional biologists since it was they who a half-century earlier had railed against the evils of "anthropomorphisms." Nothing of the sort. Without exception every letter and email that I got from biologists praised the piece, even John Alcock, whose textbook *Animal Behavior* has been the standard of the field for three decades at universities around the world and is now in its ninth edition.[1] Alcock was supportive and wrote to say that I had "found a very gentle way to educate about the evolutionary biology of emotions."

Antarctic field biologists such as Joan Larsen wrote: "All of us who spent months with penguins over many visits constantly saw behavior that as humans we would be wise to copy (but won't). I know no other word but 'love' to describe it. I have spent time in South Georgia sitting next to a wandering Albatross, who seemed so proud of its chick. I watched the courtship of those birds with fascination. And yet—the penguins stood apart in their devotion."

Another, Brian Sharp, wrote about my omissions: The colony of penguins that was filmed had declined from 6,000 pairs in the 1970s to 2,000 pairs at the time of the filming, and the reason for their decline was global warming. "Nature," I noted in my editorial, "is the greatest show on earth, and reverence for life requires acknowledging the differences between ourselves and the other animals as well as seeing our relatedness."

My implication was that love is not a mystical arrow shot from heaven. It is something as real and adaptive as a bird's beak or our hands. We have less trouble acknowledging behaviors that we attribute to "stress" across species. There is scant objection to identifying behaviors associated with elevated levels of corticosterone hormone in humans, birds, and mammals. So if penguins experience the behavioral state we label as "love," then we would predict similar hormonal and neurological changes that are diagnostic in us. On the other hand, we could not even begin to speculate what insects might feel, if they feel at all, because they have very different hormones and few behaviors that we can "read" as expressions of our emotions.

That love is a chemically induced state of mind that many animals share for the same reason, namely reproduction, did not sit well with all who read my editorial. One reader, who taught at the Chapman University Argyros School of Business and Economics and was a fellow at the Hoover Institution at Stanford University wrote: "There is now a movement afoot, led by the likes of Professor Emeritus Bernd Heinrich, to denigrate us all, to show that we aren't anything very special in the living world." It took me awhile to get a grip on this astonishing statement. I did not understand the logic of it since it makes sense only from the preposterous standpoint that Earth was made for humans alone or that we are the standard bearers of life. Does he wish to deny animals their emotions and their morals because he feels threatened, because they might make ours inferior? He went on to argue that "human beings are the only animals capable of exhibiting moral concepts about, for example, other animals or the environment or anything else for that matter." He believed that we "do very different things from other animals." Indeed! Most penguins observe parenting patterns and "moral concepts," if you will, that are distinct from those of most human beings, dogs, baboons, crows, cuckoos, birds of paradise, or malleefowl. But that is not to say that they do not exist. Even my yellow lab, Hugo, adheres to rather strict doggie moral concepts, and surprisingly some of them are similar to our own. Hugo's, for example, concerns mainly tolerance and duty to his clan, and never taking food off the table. But our morals are not the measure of other animals', nor theirs of ours. Valuing the actions or capabilities of a penguin does not denigrate those of an eagle or a cuckoo, or a human.

Other readers did not deny that a penguin's "love" might be as strong or possibly stronger than ours. But one reader commented that the use of the word "love" was "unsettling" because "Love becomes an easy and attractive tool/term for subverting social institutions and for promoting modern social identity as socially individualist. Portraying love among the penguins consecrates this tool." To this reader, love is a disorganizing force "that threatens the nuclear family," and one species' behavior can or should be seen as an exemplary model for another. Hardly.

The fact remains that the penguin's monogamy is an adaptation which is as necessary for them as polyandry is for spotted sandpipers; sexual strategy is a cog in their survival kit for reproduction in a demanding environment. Almost everything in their unique behaviors is a necessity and not an option. The eggs are laid in May and June after the birds have walked far enough onto the ice so that the spring ice breakup will not reach the young until they are old enough to handle life in the water. After the female has laid her egg it is carefully transferred onto the male's feet and tucked up into his brood pouch. He is then stuck with the egg (or vice versa) as she returns to the sea. During his continuous care of the egg during her over two-month absence, he loses about 40 percent of his body weight. It is pitch black most of the time, except possibly for the southern lights shimmering in the skies. He sleeps a great deal, but during the howling blizzards at the deep subzero temperatures the whole colony of thousands of males of which he is a part contracts and presses against each other to conserve heat. Other penguins are highly territorial, but the emperors cannot be, or they would not survive the cold. Without each other for protection against the elements, they would too quickly use up their fat reserves and then starve; they need to huddle during their long fast in the cold to conserve calories. By July, when the eggs start to hatch, the males may feed the chicks with milklike secretions produced by glands in the esophagus. But then the females start to return from the sea to bring predigested food that they can regurgitate. In the crowds of hundreds they all call for their mates, and somehow the mated pairs recognize each other, are attracted only to each other, and unite, transferring their young chick from the male's feet onto the female's and providing it with food that she has brought. He, by now emaciated, then heads to the sea to feed himself and also collect food for his chick. His return journey is made shorter by the receding ice edge, melting in the spring warmth.

Although it can be inferred that both the bonding between the mated pair and the bonding between parents and egg (and then chick) are strong within any one nesting season, that does not mean pair bonds last over subsequent seasons. Emperor penguins may live up to fifty years. Most that return to their colonies (every other or third year) end up with a different mate, probably because they must forage independently for several years after rearing their one chick, and locating the same mate is practically precluded.

I think the anthropomorphism of "love" may be justified as far as it goes, although touting any bird as a model of "family values" for humans is ill advised, unless we want to legitimize burying children in a mound of rotting vegetation and then abandoning them for life (because malleefowl do it), or sneakily dropping one's baby into another's house after killing one or all of theirs (because cowbirds and cuckoos do it). The flip side of picking one animal as an exemplar of the good and the presumed "natural" is to make it equally legitimate to substitute any other, because none is superior or inferior to another.

Animals like penguins, who require and have evolved strong mutual bonding behavior to raise their offspring, may direct that bonding in various directions. Penguins that are deprived of a mate, for example, may bond with the one they're with. For many years, two chinstrap penguins at the Central Park Zoo in Manhattan have been inseparable. They stay together, entwine their necks, and court and have sex only with each other. However, they didn't produce an egg, so instead they found a nearby rock, rolled it into their nest, and incubated it, each keeping it warm in a fold of the lower abdomen. Their zoo keeper then exchanged this rock for a fertile egg from another penguin of the colony, and the pair eagerly accepted it. The egg hatched and they became perfect parents; they spent nearly three months incubating and then raising their chick, which was named Tango by the zoo keepers. Writer Justin Richardson and artist Peter Parnell then published an illustrated nonfiction children's book, titled *And Tango Makes Three* (2005) about the pair of penguins and their chick. Curiously this book topped the American Library Association's list of banned and challenged books in 2006. It was pulled from library shelves of elementary schools in Loudon County, Virginia. Others, including some in the gay community, praised it. Why? Perhaps because the two chinstraps were males. Again, the unspoken assumption was that penguins could be construed as role models, but only those penguins who act "naturally"—like humans. Same-sex parenting is probably much more common "in the wild" than currently supposed. About 30 percent of the Laysan albatross couples on Oahu in the Hawaiian Islands, for example, are unrelated female couples.

Although that might qualify it to some folks as being "natural," it has no bearing on what is "good."[2]

The story of the penguins is a story of what is, not what one or another person may wish. Neither penguins nor other birds have "morals" in the sense of emotional pressure to act out in ways they might *not* do if left on their own. They do what they are, and are what they do. An appeal that we should be monogamous is a moral imperative that emanates precisely from those who doubt that we are what we should be. If we define ourselves as a monogamous species, then it is not surprising that some see an agenda that we all ought to live up to it. There is also a not-so-subtle undercurrent of supposition that humans represent the pinnacle of evolution, and therefore we should somehow be "better" than others. But you can't both look up to the penguins for their love, devotion, and monogamy and at the same time preach that humans are the pinnacle of evolution.

For most humans as for birds, love is not continuous nor does it last forever. What our culture would classify as "true" love typically lasts for only eighteen to thirty months. After that we either part or remain friends and stay together out of habit and perhaps mutual "nest site fidelity," as do storks and albatrosses and some other birds, or out of logical considerations related mostly to economics, comfort, and child care. This conclusion—that we are not predisposed to stay "in love" for life—is common knowledge, and it has undoubtedly been documented in cultures around the world.

Where monogamy exists in humans it is maintained at least in part by social constraints. Proof of that comes from a view of comparative culture. As anthropologist Elizabeth Marshall Thomas writes of her many years of association with the Bushmen of the Kalahari Desert of southern Africa: "A man has as many wives as he can afford, which is seldom more than two at once. But he can't marry even one, until he has proven himself as a credible provider, by killing an antelope."[3] Couldn't many species of birds be described in this way?

You don't have to be a hunter-gatherer to be polygynous. Mormons who practiced polygyny were farmers, craftsman, and businessmen. Royalty and powerful leaders, who were neither constrained economically nor by conventional mores, provide perhaps the best example of unrestrained expression of our innate desires, which appear to be polygynous; many of them were famously polygynous, if not promiscuous, keeping many wives and concubines and siring up to hundreds of children. Nowadays social constraints are used to try to enforce monogamy, or else there would not be the strong social and political censure of "scandals." They used to be private. In my "neck of the woods" in Maine, as presumably elsewhere, genealogy is not necessarily genetic. Sometimes in the old days this was even recognized in euphemistic terms such as "wood colts," which in ornithological terms might now be recognized as the result of extra-pair copulations (EPCs).

Monogamy can be and is achieved by various means. In some social animals such as wolves and people, monogamy is induced and maintained by social pressures. In a wolf pack only the dominant pair breed, although all may help to rear the pups. Any transgressions of the breeding protocol are swiftly punished. We have evolved a more democratic breeding hierarchy, and in most cases (but not all) it involves choice and equal opportunity. Yet, our biology may rebel, and in many if not most cultures we have invented marriage as a legal social construct to nurture our young. Like the wolves, we have invented social barriers and punishments to fight the mandates of our neural chemicals. As in birds, the underlying raison d'être is child-rearing.

The "traditional" human monogamous relationship has always been in part cultural, and as culture changes, so will our mores. Monogamy in humans has traditionally existed for partnerships like the raven's or like the other socially monogamous species in which the male was required to help in one vital capacity or another. The man hunted, the woman gathered; or the man went out to work, and the woman stayed and kept the home and reared the kids. With women having now entered the work force en masse, man and woman may no longer work as a team but at different business concerns. There will be separations and the need for independence because of less need for division of labor, and these will decrease the need for monogamy. Child rearing is now already partially subsidized by the local and regional governments through many educational and other programs, and it will become a lesser and eventually a moot point for marriage since many of us will have to opt out of parenthood to reduce environmental degradation.

Like many birds, however, we are flexible, and much depends on circumstances. Our idea of the right to raise many children has to change for the survival of the planet. However, we are programmed by evolution to reproduce, and like every other animal on earth we are "naturally" endowed to have as many offspring as we can, given the constraints that we face. However, nature as we know it cannot survive the onslaught of the continuing population explosion, and so given our natures and our resources and social support systems, there is little chance that our species as a whole can rely on individuals policing themselves for the common good—a good that few will recognize. However, thanks to the availability of birth control, reproduction can and has now, as in birds, become largely disconnected from loving and sex. Thanks to new laws, marriage can be and is now seen in economic terms, not only as a reproductive contract. These are positive developments, mostly because they bring about an opportunity to permit us to become closer to being truly human. They could also become not only an opportunity but also a necessity of planetary survival.

A collection from my early years in ornithology, before age ten. It includes the nest of a long-tailed titmouse that captured my imagination and the eggs of other northern European birds. 1. *Accipiter nisus*; 2. *Falco subbuteo*; 3. *Parus major*; 4. *Parus montanus*; 5. *Parus cristatus*; 6. *Parus caeruleus*; 7a&b. *Fringilla coelebs*; 8. *Carduelis chloris*; 9. *Emberiza citrinella*; 10. *Passer domesticus*; 11. *Pyrrhula pyrrhula*; 12. *Carduelis cannabina*; 13. *Turdus viscivorus*; 14a&b. *Turdus merula*; 15. *Gallinula chloropus*; 16. *Perdix perdix*; 17. *Asio otus*; 18. *Columba palumbus*; 19. *Corvus corone*; 20a&b. *Garrulus glandarius*; 21. *Sturnus vulgaris*; 22. *Pica pica*; 23. *Hirundo rustica*; 24. *Rallus aquaticus*; 25. *Troglodytes troglodytes*; 26a&b. *Erithacus rubecula*; 27. *Ficedula hypoleuca*; 28. *Phoenicurus phoenicurus*; 29. *Sylvia atricapilla*; 30. *Phylloscopus sibilatrix*.

The nuptial garb of male New World warblers (Parulidae) that I have met in Maine. Their change from nonbreeding to breeding plumage is proximately mediated by daily exposure to light (photoperiod) and testosterone.

Rose-breasted grosbeaks in various plumages, with adult breeding pair at right. First-year male after molting from first juvenile plumage is at left of the singing adult male. The pre-molt first-year male is below and the same-age female is on the right. Why is it that in geese, doves, and some other birds the young, adult, male, and female all look alike?

A pair of Canada geese nest in my beaver bog. The female (Peep), hand-reared by me, is in sleeping position while incubating on their nest on a muskrat lodge. The nest is made by pulling in whatever material she can reach while staying at that spot. The down poking out at the top of the nest rim came directly from her own breast after egg laying (when she no longer needed to insulate her belly from the cold water but instead needed to warm her eggs with her skin). There is enough to line the whole nest. Her wild mate (Pop) is guarding nearby. I easily identified him by his unique white bib. Note the injured feather on his left wing, the result of a fight with other geese.

A sapsucker male at the nest-hole entrance on an aspen tree brings in food (probably ants). Woodpeckers make a new nest hole every year. The old ones are used as nest sites by swallows (violet-green), house wrens, mountain chickadees, and mountain bluebirds. This photograph was taken at Crested Butte, Colorado.

A typical corvid (American crow) nest, built of a foundation of solid twigs broken from trees and lined with strips of bark from the wild grape. Nests are typically near the tops of tall conifers. Raven nests look nearly identical, except they are larger. Magpie nests are also covered with a roof of solid sticks, like those of the foundation here.

A winter wren and its nest built of moss and dead evergreen twigs, hidden by being tucked under the overhang of roots in a tree tip-up. The nests are lined with feathers. The largely white eggs are only faintly spotted with red-brown.

Brown creeper nests are hidden under loose, dead bark, usually of a balsam fir, in Maine. The nests are of soft plant fibers and spiderwebs on an often substantial base of dead twigs anchored by spiderwebs. A cross section through the tree just above the level of the nest is at right.

An American woodcock female incubates and hides her eggs. She allowed me to photograph even though I was directly over her. She did not blink.

The same woodcock nest without the bird covering the eggs. A normal clutch has four eggs. The ferns indicate that this is late in the season, so this is probably a replacement clutch because the first clutch is laid in late April, long before ferns are up.

Looking straight down on the nest of a dunlin (sandpiper) on the Alaskan tundra.

The dunlin nest is revealed by teasing apart the natural cover of sparse dry sedge. Note that it is too early in the season for new leaves, although one shoot is coming through the soil in the lower right corner of the photograph and several to the left of the nest.

Looking down on an ovenbird's nest on the forest floor. The nest is in the center of the photograph under the twig. It has a roof and is for all practical purposes invisible from this angle.

A mourning dove incubates her (two) white eggs on her nest in an apple tree near my house. The nest is nearly invisible except for the clue of a few loose, dry twigs, as long as the bird remains in place.

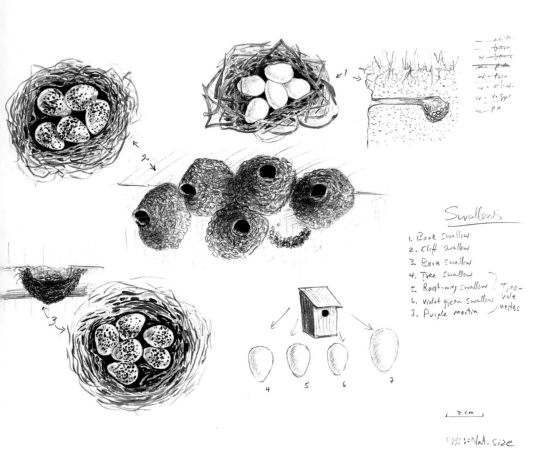

North American swallow nests and eggs. Those species nesting in tree cavities and those making burrows in sand for nesting have white eggs. Those making mud retorts for their nests (of grass and feathers) have spotted eggs.

American robins typically nest in a solid tree crotch or crevice against solid substrate. In Vermont, eggs in nests such as this one are usually raided. Others that are built on beams or other human-built structures usually fledge young. Note the scraggly outer appearance of the nest. Inside is a solid (if dry) mud cup lined with fine dry grass that cradles four blue eggs.

Nest of a yellow-olive flycatcher next to a nest of wasps *(Parachartergus fraternus)* and possibly two fake or partial nests. This species commonly nests near wasp nests and thereby gains protection from some predators.

A red-eyed vireo incubates on its hanging nest in a sugar maple tree. It is shielded from above by leaves. The bird does not nest until the leaves emerge.

The vireo and white-faced hornet nests.

The northern oriole nest is a bag with an opening at the top. It is woven from long fibers pulled from dead milkweed stalks. Nests typically hang from the tips of branches, far from the trunk.

Winter wren nests in New England commonly sit near the ground, tucked into crevices among roots and soil of tree tip-ups. Most are difficult to spot except from up close unless you know what to look for. Note the lateral entrance hole *(center)*.

The gray catbird's nest is a cup, like the robin's nest and like most bird nests. But the materials used for it and the resulting features vary enormously. The nest of this species is lined with fine roots garnished with strips of wild grape bark. It sits on and is surrounded by a loose framework of scraggly twigs. The nest is built low to the ground in dense shrubbery.

Here I am being dive-bombed and shat upon by a pair of Arctic terns while photographing their nest on a sandbar of the Noatak River in Alaska. The camera is pointing directly at the nest. (Photograph by F. Daniel Vogt.)

The resulting photograph of the Arctic tern eggs in Alaska. The eggs are laid onto bare ground without any visible nest, and they are camouflaged among the stones and gravel. Arctic tern nests examined among vegetation on the Isles of Shoals off the coast of Maine were lighter in color and had more green.

A mourning dove sits on its minimal nest containing, as for most doves and pigeons, two white eggs. The pair take turns incubating but only change in early morning and evening, thus the nest and its contents remain hidden and are not divulged by activity near it.

This Canada goose was conspicuous while incubating on the top of a beaver lodge, but such commonly used sites in Vermont provide safety from the surrounding water. This particular nest failed, however, because the beavers chewed up from the inside of the lodge and the nest bottom collapsed down into it. Canada geese also use old nests of herons and raptors.

*Above:* A common loon incubates on its nest at the edge of a pond. The sharp "fish spear" bill of this large bird is probably an effective defense against many potential predators. *Below:* This nest failed because it flooded in record rains that raised the water level.

The nests of white-fronted geese near Barrow, Alaska, are highly visible. Note the early nesting time; the snow has not yet melted and more will fall. Geese line their nests with down from the female's own breast. Here the wind has scattered some of the loose down. Although very visible, the nest is strongly defended by the gander. It snowed a few days after the photograph was taken, and tracks in the snow revealed that an Arctic fox had circled the nest but had been repelled by the geese.

Great blue heron colony in a bog created by a beaver dam that has killed the trees. The nests are highly visible but relatively inaccessible because of their height and the water. They are often recycled for use by ospreys, red-tailed hawks, and Canada geese.

The nest of the great crested flycatcher often contains a snakeskin. This one also has owl and goose feathers. The nest was in a bird box.

The nests of the broad-winged hawk are lined with bark chips and typically also contain a sprig of green hemlock. The nest is almost always in a triple crotch of a large broadleaf tree.

The nest of the snow bunting is a thick, well-insulated structure of dried grass lined with feathers. This one was located in a crevice in a jumble of rocks on Ellesmere Island. It was later abandoned by the birds when it was taken over by a bumblebee queen for starting her colony.

A whimbrel nest along the Noatak River, Alaska. Note the pyriform shape of the four eggs. The pointed ends are directed down and toward the center of the nest scrape, as is typical in all northern shorebirds. This configuration maximizes the brood space of the incubating female so that the largest possible eggs are laid (and young hatched). High Arctic shorebirds' young are precocial on hatching and have only a short seasonal time window to grow up and be ready to migrate before freeze-up.

A spotted sandpiper nest on a sandbank along the Huntington River near Richmond, Vermont. Most shorebird eggs will be incubated by females, but these could be incubated either by the female or by a male. If the female finds a second male, then she will leave the first clutch for him to incubate. She will lay a second clutch with the second male and incubate those eggs herself, thus permitting her to produce eight young in the season even though she could only incubate four eggs at a time.

A full clutch of mourning dove eggs. These birds, like hummingbirds, belong to a group with a generally more tropical distribution. Doves lay few eggs at a time but may breed year-round and produce many consecutive clutches. Such eggs are seldom exposed to the light of day because one of the two parents covers them continuously. Hummingbirds probably gain much more protection from predators than doves because their walnut-sized nest resembles a knot of a tree limb.

A clutch of mallard duck egg. Ducks typically lay many egg per clutch, and all have to hatc at the same time. This mean incubation cannot start unt almost two weeks after the fir egg is laid. In the meantim the nest must remain hidden The female covers the eggs b scraping nearby debris ove them before she leaves afte laying each egg each day. (Her the hen left in a hurry and have pulled the nest coverin apart to take the photograph Immediately after they hatc the female leads the young t water where they can fee themselves.

Herring gull clutches are typically only two to three eggs. Although the young are physically precocial, they must be fed by the parents, unlike sandpipers, ducks, geese, and grouse. Parents are limited in the amount of food they are able to find and hence the number of young they can have. Pigment is applied onto the egg surface at different times as the egg passes through the oviduct, resulting in depth of coloration through the partially transparent calcium layers.

Eggshells can be thick and hard. Part of the solution to the problem of the chick's escape from the egg is accomplished with a tool that the young are born with (but quickly lose)—the egg tooth. It is visible on top of the bill's tip of this just-hatched herring gull chick.

A hatching great horned o[w]
chick. When dry it will be co[v]
ered with white fluff yet still d[e]
pendent on the mother to bro[od]
it. The intact egg shows a bul[ge]
where the second chick is abo[ut]
to crack through the shell. [All]
owls' eggs are round like thes[e.]
The male parent has alrea[dy]
provided a vole, which the [fe]
male will tear into tiny bits f[or]
the chick, who will pick the[m]
from the parent's beak. Thu[s,]
like most raptors, the owls a[re]
more precocial than songbir[ds]
but less so than sandpipers. T[he]
object above the vole is a reg[ur]
gitated pellet.

A minimal nest scrape in the open holds two of the eventual four in a clutch of pectoral sandpiper eggs. The eggs lie uncovered for three days after the first egg is laid before incubation starts. Note the disruptive coloration that breaks up the large egg's outline.

This red-winged blackbird nest is built into the center of a dense cluster of sedge surrounded by water. Note the odd and intricate markings of the eggs, which are individually distinct. They differ even more from one bird to another.

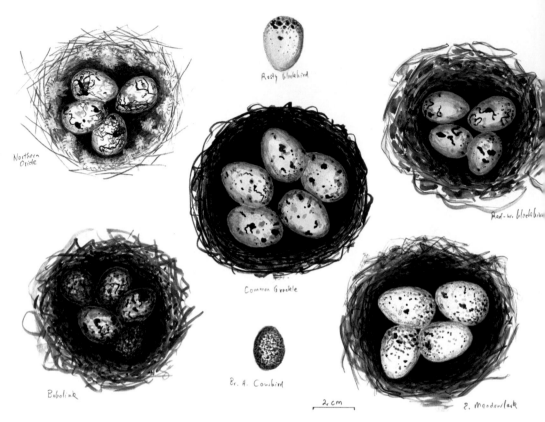

The eggs of various blackbirds and their kin. Most families of birds have egg color patterns that vary according to a relatively consistent scheme, but those of the icterids, as indicated here by seven species, vary greatly: *(above center)* rusty blackbird; *(above left)* northern oriole; *(center)* common grackle; *(above right)* red-winged blackbird; *(below left)* bobolink; *(below center)* brown-headed cowbird; *(below right)* eastern meadowlark.

Just-hatched altricial young—here, cedar waxwings—can do little more than raise their heads and open their mouths for food to drop in whenever they sense a vibration, which is likely to be a parent landing at the nest. They can also rear their hind end to present a fecal droplet, usually in response to being fed, so that the parent can accomplish both feeding and nest sanitation in one visit.

Precocial young of the Canada goose are able to swim, run, dive, swim under water, stretch, peck at food, stand and flap their wing stubs, and feed themselves the first day after hatching.

A day-old killdeer. On hearing the warning calls from parents, the young hug the ground and freeze. Precocial birds of various species typically have intricate and beautiful species-specific down garb that camouflages them much as the eggshells camouflaged the embryos.

One- or two-day-old herring gull chicks. Note the eggshell by the nest that has not yet been removed but will be. In the meantime, a parent is always on guard next to the young, which will leave the nest in several days and only gradually wander from the nest vicinity. Nesting nearby, black-backed gulls will eat young herring gulls if they can catch them.

Canada geese typically lay four to five eggs. Unlike in ducks, the males stay with the families and act as bodyguards. Families gain additional protection by joining up into groups. This one with twelve young is likely a composite from at least three nests.

A mourning dove nest with the young. The adult bird did not flush, nor ever move or blink, even as I climbed the tree and photographed it from a half-meter away. Note the fecal accumulation around the nest edge. Waste removal would require leaving the nest unattended, and this nest was attended by an adult without any apparent daytime breaks from when the eggs were laid until the young fledged.

North American goldfinch young are ready to fledge in late August. They feed on thistle seed. The young defecated over the nest edge, and the nest rim is ringed with feces. These are the latest-nesting birds in New England, and one of the few that leave their young's wastes. Many other birds eat the wastes. Is their diet a factor?

# 5

# FINE-TUNING
# NESTING TIME

IT IS THE LAST WEEK OF MARCH 2007,

and I'm at my Raven Hill Camp in Maine. Temperatures on the

minimum-registering thermometer show that it had recently

been −29 °F. Despite the cold and the snow, it's a great year

for birds—flocks of chickadees feed on the spruce, larch, pine,

and hemlock seeds that have been scattered onto the surface of

the snow by the red-breasted nuthatches, pine siskins, and pur-

ple finches and crossbills when they pick at the cones in the treetops. White-winged and some red crossbills have made a rare appearance this year because there is an abundant seed crop.

Unlike most years, I met the crossbills everywhere on my drive through Vermont, New Hampshire, and then into western Maine. I stopped at several places to listen to the incessant singing of the males attired in their purple nuptial garb. They perched on the tips of larches and spruces, and groups of them, taking singing breaks, flew onto the road, picking at salt or gravel. Crossbill roadkill was frequent, and I picked up one male and opened him. Both his testes were enlarged to the size of kidney beans. The crossbills were breeding now because they were stimulated to reproduce by the now-ample food supply. I reached camp walking through deep snow, surprised that there were no crossbills at this particular area, where they had been plentiful several years earlier. They may, like some other birds, clump up locally to breed. Have they evolved to seek company because it facilitates mate finding and extra-pair copulations, or do they use others as cues for finding food or suitable habitat?

I woke early the next morning in my log cabin, and as I was making my coffee I heard the local raven pair making a commotion of rasping angry-sounding calls at their traditional nest site in the nearby pines. Odd, I thought, because at this time ravens normally incubate their four to six blue-green, black, and gray-blotched and -spotted eggs and they are usually quiet, almost secretive. I rushed over to the nest to see if there was a problem. But when I got near the pine tree with their great nest just under the canopy of thick branches about fifty feet up, the ravens were gone. Instead, a great horned owl flew from their nest tree. I had climbed almost all the way up to the raven nest when a second owl, presumably the incubating female, suddenly flew from the nest and hooted from the nearby woods. When I reached the nest and maneuvered around the edge to be able to look in I saw that it contained the fresh lining of deer fur and bark that the ravens typically use, but it cradled two large round white owl eggs. On April 8, after another two feet of snow had fallen, the ravens were again in an altercation with the owls at the same nest site. This time the owl flew off the nest even before I stood under it. Except for a crater in the center, the nest perimeter was heaped with a foot-high wall of snow, with an entrance hole to the eggs inside on the side.

Ravens, great horned owls, and crossbills are the early nesters in these woods. The ravens now feed on coyote- and winter-kill, and the owls hunt hares and grouse. Most of the birds would not be back here until mid-May, but some other resident birds had also prepared to nest in what appeared to be winter. On March 24 I had heard a pileated woodpecker for hours in the morning making its high-pitched rapid staccato calling from the woods near our house. The continuous calling was so unusual that I went out to investigate, and found a standing poplar tree that was broken off halfway up. A few feet below the break was a large round hole hammered into the bole, and the deep snow below was strewn with fresh yellow wood chips. In the afternoon the woodpecker was drumming loudly on resonant dry wood about 400 m from the nest site, and it kept up the drumming display for at least an hour. When I later went to the poplar stub to investigate, I heard hollow, steady tapping; one of the pair was excavating the cavity in preparation for eggs and young.

I knew it was spring when within another two days I heard the first song sparrows sing and mourning doves coo. Crows and blue jays called during the day and the woodcock sky-danced in the evening. I heard the loud honking of the first pair of Canada geese approaching from the distance. There was so far no open water, and they landed on the still thick ice of our beaver pond. After walking around for about a half-hour, they flew off. They will now be back almost every day to lay their claim to the whole bog and particularly to the prime nesting site, a beaver lodge that will soon be surrounded by water.

The early bird may not always get a worm, but it does get first dibs to a territory. In most birds the males, generally the older ones, arrive first. They get the best territories either by starting their migration earlier or by migrating more rapidly, but they may also pay the ultimate price if they are hit by bad weather on the way or on their arrival. The geese that had once nested here will be more willing to fight to stay than others who come but have not yet had a chance to use and value it. These early birds, although perhaps not consciously anticipating food, will find tender green shoots when their goslings hatch.

The crossbills nesting in winter while feeding on spruce and larch seeds and the goldfinches breeding in August when the thistles come into seed are "exceptions" to the rule of spring being most birds' nesting time. But they prove the rule that breeding time is feeding time. Most other birds can reliably expect food in the summer, so cues for breeding time may not always be food, as such. Photoperiod is also a common cue to get ready to breed, because lengthening days are correlated with increasing food supply.

Open ground is one of the first sites where food is available to robins, song sparrows, crows, woodcock, and grackles. Those who feed there are among the earliest migrants to breed after the snow melts. Great blue herons, ducks, geese, mergansers, ospreys, and belted kingfishers arrive after the ice leaves the lakes and streams and feed in or at open water. They are followed by birds catching flying insects, which include various flycatchers. Finally, just as the trees leaf out and caterpillars and other insects arrive, come the orioles, tanagers, many warblers, and vireos.

As much as the birds' arrivals are correlated with the availability of food, birds migrate from such long distances and travel too fast to be able to predict the actual amount of food they will find on the ground or in the trees. They can only predict the probability of finding food and fix that probability to a date on their internal calendar.

The birds that I don't see in our woods until late May or even June may seem to be laggards, but they have been preparing to nest for a long time already. By February, all over the northern hemisphere birds are getting ready to return to their nesting grounds. Some are already on their way.

They often face storms and starvation. In the first week of February 2008, a continent-wide weather system of tornadoes and torrential rains dumped a foot or two of snow all over New England and I wondered if birds got blown off course. As I got up on the morning of the 9th and looked out under the still-overcast sky that showed the first hints of clearing and blue, I saw a flock of over a hundred redpolls swirling off a birch tree like leaves blown by the wind. It will be more than a month before they and the pine grosbeaks and Bohemian waxwings return to and disperse on their breeding grounds in the Arctic tundra. Then I thought I was "seeing something"—could those shapes perched in the top of a poplar tree be robins? I look again: definitely bigger than the pine grosbeaks. I grab my binoculars and take a second look: yes—orange breasts. I have never seen them back this early, and I run out to confirm and to wonder. Indeed robins—eleven of them, and among them, oddly out of place, is also a cedar waxwing. A minute later they all flew off, but I then heard the sweet two-note song of a tufted titmouse (who stays all winter subsidized by our bird feeder).

I would not hear the first song of the robin until the first week in March. At the same time, black-capped chickadees burst forth in an exchange of "dee-dahs," hairy and downy woodpeckers drum, and overhead, scattered phalanxes of crows are heading north on steady wing beats, while others caw excitedly from the pines. The high thin call of the blue jays and the staccato chatter of a white-breasted nuthatch resound from the woods, where the snow crust is hard enough to walk on. There are, here and there, tiny patches of bare ground along roadside banks and in the fields, and any day now one expects to hear the first "oog-la-ees" of the redwings in the morning and the first woodcock "peents" in the evening.

Birds' nesting time in the north is brief, and in many species it is on a precisely choreographed schedule set to internally timed annual rhythms.[1] Many northern beetles and moths require a year, and sometimes several years, to develop from egg to adult, but in the space of as little as a month most northern woodland birds court, mate, build a nest, and raise usually not just one but often two or three broods to adulthood. In order to start this process, they first have to migrate a distance of several thousand miles from the warm tropical areas where they live for most of the year. They must also fit in a molt, or sometimes two, in order to have the appropriate breeding garb and flight equipment to negotiate their long-distance journeys. They must fuel up for their magnificent ultra-long-distance journeys at the appropriate times by feeding frenzies in which they put on the massive amounts of fat needed to power them for up to three or four days of nonstop flight. Because of their constant presence, I am perhaps most impressed by the delicate and colorful motley of wood warblers (Parulidae) that are particularly diagnostic of these north woods.

In June I am in almost continuous daily and hourly if not minute-by-minute contact with any one of some twenty species of warblers. These birds are rightly called the "jewels" of the northern forest, and among themselves they carve up the northern forest into diverse ecological niches that serve as their homes and nesting places. One species, the blackpoll warbler, is a denizen of the northern spruce-fir forests that stretch across northern North America from the East Coast, across Canada, and to the west and up into Alaska. In New York and New England it breeds only near the tops of spruce-clad mountaintops, where in the summer you hear its very high-pitched, wheezy and thin repetitive "tsi-tsi-" song that is to us as nondescript as its black-and-white plumage and its characteristic black cap. Where I live in Vermont I occasionally hear this wheezy, high-pitched song of blackpolls in late May, from migrants passing through on their way north and west to their breeding grounds.[2] Nests will be built by the females, clutches of five to six eggs laid, and the young fed caterpillars and other insects by both parents for about ten days. After this breeding is over (the pair may occasionally build another nest and raise another brood), the families split up, and both the young and the adults return to South America. The migratory timing, direction, and navigation by star compass and magnetic and visual orientation involving ultraviolet light are behaviors encoded in their genes; they leave and return "knowing" when and how to go and how to prepare.

By early September both the young of the year and the adults from Alaska will have flown 3,000 miles to rejoin and mingle with those of their species from the rest of the continent that have gathered in the east near the Atlantic. (Their spring migratory route is different than their route south—on their return the birds come more directly north rather than retracing their trip to the eastern seaboard). Until then they all will have fed themselves and their young primarily on insects. Next on their schedule is a huge feeding binge on berries. The birds, a mere eleven or twelve grams, fatten up to become little butterballs of about twice the body weight they had while breeding and when reaching adulthood after leaving the nest. This amount of fat, burned at 0.06 percent body weight per hour in flight, is enough for about ninety hours of continuous flight. They carry barely enough fat required to fuel their flight from northeastern North America to South America. For the birds' awe-inspiring nonstop flight—eighty to ninety hours over open ocean and all the way to South America—they have little fat to spare, because they are forced to fly during the day as well as at night (most other small birds migrate only at night and feed in the daytime).

The signal for departing out over the open ocean is usually a cold front from the north that helps them with a tailwind. Radar data show that after they leave the North American landmass they rise to about a half-mile in altitude. They attain about a mile in altitude over Bermuda, but later rise up to three or four miles to avoid the southern headwinds at the lower latitudes. Finally, in the last few hours before reaching the South American coast, they descend to 2,500 feet before landing. All of their three- to four-day nonstop flight from their nesting grounds is fuelled by their "on-board" fat; they do not stop to eat, drink, sleep, or rest.[3] Presumably their spring return flight to their nesting areas is no less remarkable.

The cross- and intercontinental sojourners nesting in North America, Europe, and Asia are on relatively strict schedules that are genetically inherited and thus internally generated.[4] Rapid evolution has produced geographically distinct populations in some of the species of songbirds, and although they continue their fixed schedules of behavior and physiology under constant conditions in the laboratory, they are also synchronized to local photoperiod, which acts like a cue—much as we might set our wristwatch to local time.[5]

Birds breeding in the short northern summer raise usually one but sometimes up to three broods to adulthood, and have them all ready to leave by fall. If we add the time to build nests (as fast as three to four days in some songbirds with simple nests, ten to twelve days in swallows, and twelve to twenty days in some who build relatively elaborate nests), and lay the clutch (two to three days per egg in large birds and one per day for up to twelve days in small birds), then that leaves little time for a second brood. Even some of the small birds (such as tree swallows, woodpeckers, penduline tits) don't breed twice in a summer. Some, as indicated previously, manage to do more by having two or more clutches going simultaneously through adjustments of their breeding systems that involve polygyny and polyandry.

Tropical birds have very different life-history patterns, although

it is not always clear if that is because they are different birds, or because their environment is different. One of the well-studied comparisons within the same genus, *Sylvia* (the Old World warblers),
which occurs both in the temperate north and all the way to southern
Africa, was conducted by a team of four biologists, Hans-Christian
Schaefer, George W. Eshiamwata, Fred B. Munyekenye, and Katrin
Bohning-Gaese. Bringing expertise from both Gutenberg University
in Mainz, Germany, and the National Museum in Kenya, they set
out to compare the life histories of two African warblers, and then to
compare those with the already well-known pattern of their northern
congeners.[6] As is typical of other tropical birds, the two southern species had clutches of only two, rather than the five or six expected in
Finland, but they had three to seven clutches per year rather than the
one or two per year in Finland. However, on average the two south-
erners only fledged 2.3 and 1.4 young per year, in part due to longer
incubation durations and nestling periods as well as high predation
rates. In great contrast to northern warblers, in which the young are
independent several days after leaving the nest, these tropical warblers required more than a month and up to two months to become
independent (no longer fed by parents), and even after that they continued to stay in the parents' territories. These results are in agree-
ment with previous comparisons of distantly related birds in other
tropical areas. The conclusion is that nest predation is higher in the
tropics and the extended care of juvenile birds in the tropics may be
a compensation resulting from the heavy nest predation, helping ensure the longevity of those who do survive the nestling period. It has
been speculated that it would likely be counter-productive for tropical birds to rear larger clutches, because it would increase nest visits
by the parents and give away the nest location to predators, and thus
potentially increase predation instead.[7]

Only small birds manage to fit several broods into each summer, thus the number of broods per year depends on the latitude

where they live. In India the house sparrow nests year-round and produces seven clutches per year, while in central Europe it nests four times per year, but north of the Arctic Circle only once. In part, larger clutch size compensates for fewer clutches, presumably because longer days in the north allow more foraging time per day in the summer. If the nesting attempts of large birds are disrupted and fail, then there is little or no time for re-nesting in a north-temperate season. In contrast, the many small passerines whose nesting attempts are disrupted will attempt to re-nest within several days. In Vermont some savannah sparrows (whose nests are often destroyed when hay is harvested) may attempt to re-nest up to six times in a season before they finally bring off a clutch of young (Noah Perlut, pers. comm.).

Many birds are either strictly seasonal in their behaviors and times of reproduction, or reproduction is timed to coincide with specific signals from the environment. Additionally reproduction in those who are monogamous is also synchronized as one mate adjusts its reproductive schedule to the other. And a large part of this timing is flexible and more precise than the season as such. It can be switched on and off, depending on immediate circumstances. We know that hormones affect behavior; male birds typically have high levels of circulating testosterone during the breeding season (their gonads are then greatly enlarged), which developmentally affects their breeding garb, and behaviorally their singing and their territorial and mating behaviors. Females experimentally treated with testosterone become less attractive to males as extrapair partners.[8] The interesting question becomes: How is the adaptive suite of on-off behaviors regulating reproduction achieved? What are the "switches" and how are they activated to initiate and stop nesting season and fine-tune the behavior within the season?

*Nesting behavior and hormones.* Pairing up and nesting is orchestrated by a dance of a half-dozen hormones in constant interaction and with feedback from sensory stimuli from the environment. They include follicle-stimulating hormone and luteinizing hormone from the posterior pituitary, which stimulate the ovaries to produce estrogen and progesterone. Estrogen helps develop the oviducts and stimulate nest building behavior, while progesterone stimulates egg laying. The tactile sensory stimulation of the eggs releases prolactin, which promotes the incubation behavior.

My first, and memorable, introduction to this synchronization was during a seminar I attended at UCLA in the early 1960s by the late Daniel Lehrman, then at Rutgers University, who spoke of his experimental work on the reproductive physiology of ring doves.[9] The reproduction of doves and pigeons may seem routine and hence boring because it occurs so regularly and faultlessly. But Lehrman looked at their reproduction from the outside in and the inside out, and showed it to be a process with universal implications (which I presume have not changed fundamentally in the last half-century). These doves can nest year-round in the laboratory, and so he wondered why a bird at one time has greatly enlarged gonads that pump out sex hormones, and at other times only tiny shriveled gonads and no interest in reproduction. What turns the bird on and why and with what or whom? In the usual behavioral reproductive cycle of the dove, a male and female meet, court, build a nest, mate, lay eggs, incubate, and both produce "crop milk" (a secretion of sloughed-off cells from the crop), feeding their two young with it. All of this is, on the surface, "behavior." But each step occurs only at a specific time, and it is orchestrated inside, by physiology.

Reproductive physiology that drives behavior is commonly stimulated by signals from the environment. The sensory stimuli affect the brain, which in turn affects the rest of the body through hormones. Lehrman had taken blood samples and followed the changing profiles of the hormones through the breeding cycle with its corresponding physiological changes and behavior. Estrogen is responsible for the growth of the ovaries and their production of eggs in the female, and testosterone turns on the males' courting and mating behavior. As the male's testosterone level rises, he is primed to court a female if he sees one. She, stimulated by seeing the courting display, is then induced to secrete gonadotropins (hormones that stimulate the gonads to grow and produce their hormones, estrogen and progesterone). The estrogen stimulates nest-building behavior, ovulation, and receptivity to mating, and the progesterone primes her to incubate at the sight of eggs. The sensory stimulation of incubation in both her and her mate then induce prolactin production, which then induces and maintains the bird's production of crop milk. The young, when they arrive, are induced to be fed with the presence of crop milk. Thus, at each stage of the breeding cycle the doves are primed by appropriate hormones to respond to specific relevant sensory stimuli, which then get them ready for the next step in the breeding cycle.

These broad outlines operate within other cycles. First, long photoperiod stimulates the brain of the male to produce a gonad-stimulating hormone, which acts on his testes to make them grow large and synthesize testosterone. The testosterone then affects the male's brain neurons, making him coo to a nearby female. In turn, the female's brain is stimulated and she starts to produce corticotropic hormone, which stimulates her ovaries to grow and produce estrogens. But only up to a point. The male primes her reproductive drive, but that alone does not yet release the nesting instinct. For that she needs the presence of not only an uncastrated male but also materials with which to build a nest.

To me it seemed magical almost beyond belief that the presence of twigs and the sight and sound of a cooing male could cause the physiological changes in the body of a bird to induce her to make an egg. Sheer magic! But of course the sensory stimuli required to orchestrate the nesting of the dove is specific. It involves the sight of both the courting male and the little sticks. A mere male, as such, doesn't do it. It has to be a cooing male—one that has intact testes. If all of the proper visual and aural stimuli are in order, then the pair mates, and the female lays the two white eggs typical of almost all doves and pigeons.

After the eggs are laid, the dance of the hormones continues, because then the feel of the eggs on the bird's breast stimulates prolactin production, which induces further incubation. Higher amounts of the same hormone act later on in both birds of the pair, enlarging glands in the throat that then secrete the white "pigeon milk" that is fed to the young squabs soon after they hatch. Lehrman produced what seemed to me an incomprehensible map with arrows going in all directions, indicating what affected what in chains and circles and alternate routes that drove home the enduring truth of it all: everything is connected to everything else, and reproduction involves stimuli that may be arbitrary for us but are specific, even essential, to a particular species. These stimuli are filtered through the brain, where they turn on the endocrine system, which releases the hormones that affect both behavior and growth of ovaries and eggs, testes and sperm. Simple.

The hormones in the dove are the same ones that appear in all birds and in mammals, where they induce some similar but also many very different effects. Yet the same principles apply, even though the stimuli and the targets of the steroid hormones and the prolactin differ decidedly among different animals. Most birds, for example, have crops but do not have glands that can be activated to produce "milk," although the same hormone is responsible for milk production or release in mammals. In various birds prolactin also results in fattening and migratory restlessness. In humans prolactin functions in bonding, presumably by the usual mechanism of the hormone binding to receptors of the appropriate neurons. In birds it acts similarly, although in this case it could be bonding to the eggs and then the young. High levels of it are associated with paternal and maternal helping behavior in cooperative breeding in Harris's hawks.[10] Nest building, which is stimulated by estrogen in pigeons and doves, rats and rabbits, is stimulated by progesterone instead in mice and hamsters.

One may wonder why monogamous birds that are already paired bother to court each other. The experiments by Lehrman suggest that courting is not just about getting a mate. It functions also to synchronize the breeding activity of an existing pair. It starts the cascade of physiological responses that result in nest-building, mating, egg-laying, incubation, and then the process of caring for the young. The question becomes not just how the birds synchronize their nesting to the environment, but also how they get others to synchronize to them.

6

STRUTTING

THEIR STUFF

IT IS THE END OF APRIL, 2008. TWO DAYS
ago on my drive from Vermont to camp in Maine I listened to
a story on the radio about a percussion musician who made
his mark using recycled junk as instruments. Different timbres
and rhythms were produced by striking tin plates, garbage cans,
and so on. Music is noise if it is not pleasing. But it all depends
on the competition. This was music for some. That evening at
camp in the Maine woods I heard a far more inspiring composi-

tion "sung" by a woodcock. It had a vocal component with an instrumental accompaniment. The latter is produced by specially modified wing feathers that are bent to produce a flutelike sound as they vibrate in concert with the wingbeats. The music was performed during a spectacular aerial dance in the moonlight, about fifty to over a hundred meters above the forest canopy. The next morning I woke to a clear and beautiful day, and almost the first thing that I heard was the drumming of a sapsucker. Of all of the endless dead twigs, branches, logs, and trees that he could have used, he apparently chose to make his music on what would make the most noise—an aluminum ladder standing next to my cabin. With it he made a cheery, high-toned sound that was even more impressive than pounding on a pie plate. He produced a rhythm that went something like this: rat. tat.tat—tat—tat—tat—tat.tat.tat—TAT, and many variations thereof. Some of them were so fast and varied that I could not reproduce them. I went out and tried to photograph the performer because he was so gaudily decked out in black, white, crimson, and lemon yellow, but he was camera shy and flew off into the woods. He returned to resume his performance soon after I was back inside. In a short while I heard many more sapsuckers performing. These had only dried wood available and did not sound nearly as inspired. At times they stopped drumming and I heard them "talking" loudly. They were drawing attention to themselves in sometimes less musical rasping series of screeches that had upward inflections at the end, as if for emphasis. Then they resumed their drumming. I heard them from all sides. One produced high sharp sounds, as by drumming on a thin, dry twig. Another made lower sounds as on a long, dry plank, and still another made dull sounds as if from a log. The birds were trying out different media and then seemed to form preferences; back in Vermont I had attended another of their concerts, at which one of the three sapsuckers there came repeatedly to drum on a metal stove pipe.

We tout our noisemaking as music because it is innovative, and would question whether birds make "music" because it consists of specific choreographies. But on the rare occasions that I can bear to change stations, most of the music that I hear also sounds as if it were part of a genre. Music, some would say, is made for enjoyment, and the bird's is made for an ulterior motive of sexual attraction, but I would bet my bottom dollar that the most immediate reason why they do it is because they like to; i.e., for enjoyment. Their message, to them unknown, but an adaptive one for which it has been selected, conveys: "*Notice*—here I am, and I can do something as good or better than anyone else." Actually, I think that's exactly what I'm hearing from human performers, too. Indeed, musical performance in humans is likely a selected trait that has the same utility as the sounds made by crickets, frogs, birds, humpback whales, hammerhead bats, and howler monkeys. Superior song is a privilege of the few that have mastered it best, and it is evaluated. It had and still has consequences for reproduction.

On March 18, 2009, I woke to a dark, very windy morning that was threatening rain. Most of the birds were quiet, as they often are in bad weather, but this morning I heard the first song sparrow of this spring. I presume the bird came in aided by the wind during the night. Again and again he burst forth in song, perched in the top of a still-bare white birch tree next to our garden and close to the bog. I have no ability to remember the individual songs of this virtuoso, but I tried to fix each song in my mind long enough to be able to compare it with the next one, which came about four or five seconds later. Almost invariably the second song was different from the one before. The sparrow stayed on the same branch, first facing in one direction while singing, then facing in another direction; there were no answering songs. After singing for about an hour and a half, he flew to the ground in the bog to some of the already snow-free spots before singing from the top of a low shrub. His songs were now less varied, except for their end flourishes. By 1 PM he sang less vigorously and only sporadically.

On April 15, 2009, the dawn promised a gorgeous day; the sky was clear, it was still, and the temperature was about −4 °C. I was up long before the sun came up over our ridge to cast its golden light onto the pines on the other side of the beaver pond. There was then a flurry of birdsong: robins, mourning doves, the phoebe, tufted titmice, all chimed in along with a pair of dueling chickadees, goldfinches, rasping screeches and staccato drumming of sapsuckers, and screams of blue jays. Some of the migrants from the north had not yet left—pine siskins and a fox sparrow acted as though they were courting. In the bog I heard the wonderful quacking of a mallard and the wild calls of the geese above the background babble of grackles and red-winged blackbirds. But my main focus was on a song sparrow near our shed. He sang incessantly, about six songs per minute. Many of the songs were of the same type, one after another with a slightly different ending, but then suddenly he would throw in a different type to enrich his repertoire. There were no challengers, but I did see another sparrow around, who stayed silent.

I was perched comfortably on a limb in a sprawling pine tree at the edge of our bog on April 27, 2007 when I noticed a song sparrow below me come up out of the dense mat of grass and sedge among willow bushes. He periodically flew to the top of a bush and sang. I also heard a soft, almost surreptitious song from the brush close by. Right after that I saw a vigorous chase that ended as one bird pounced on another that had apparently been there all along in hiding. In their struggle both landed on the water next to a beaver dam, where, as if frozen, they stayed still a few moments with their wings spread out onto the water. After a pause they suddenly separated as one returned to the tangle of brush while the other flew up onto the tip top of a bush and sang loudly. Both of these song sparrows had ignored a nearby swamp sparrow, which was directly under my perch in the pine the whole time. To each bird the other's song was probably no more than noise. The songs of their own species, however, have meanings that ultimately relate to parenting.

Music is a big part of their battle for mates and territory, and it makes that of the "dueling banjoes" in the movie *Deliverance* look tame. The naturalist John Burroughs wrote about the song sparrow in 1877, noting that "it displays a more marked individuality in its song than any bird with which I am acquainted. Birds of the same species generally all sing alike but I have observed numerous song sparrows with songs peculiarly their own."[1]

Individual song sparrows' singing (and the variety of songs by the same individual), courting, and nesting behaviors were later studied for many years by Margaret Morse Nice, a pioneer behavioral ecologist of the 1930s, who said, "The study of animal behavior is the only and ultimate source of understanding ourselves." She kept track of the travails of sixty-nine banded pairs of song sparrows with adjacent territories by her home in Ohio.[2] She noted that songs varied from one individual to another as our faces do, even though the song types are clearly recognizable as those of the same species. Furthermore, most individuals had not just one but a repertoire of several songs.

*Song.* Much research effort has been expended to try to understand the significance of song repertoires in birds, and in the song sparrows' repertoire of songs, one of the few species that have been studied in great detail, current themes include deciphering why they have them and how they use them in "honesty in signaling."[3] The issue boils down to the fact that the vigor of a song is in itself a signal that functions in attracting mates and in keeping intruders of the territory at bay so that fights are avoided. Rivals will generally hesitate to start a fight they know they cannot win. But the question is: If the rivals for a mate or a territory are evenly matched, then the one who exaggerates his power should win by default as the other surrenders without a fight. So why, then, don't birds exaggerate their vigor to get a mate or to intimidate a rival? The seemingly easy answer is because it is not likely to evolve as a stable evolutionary strategy: if all exaggerated the signal, then it would become meaningless and could be ignored. A lion's roar really does most of the time need to have fangs behind it to be feared. For any potential signal to actually function as one, it must on average be backed up. Studies with song sparrows' vocalizations in the context of their natural history have parsed out how such honesty in signaling is achieved.

The song repertoire of song sparrows varies from five in some individuals to up to fifteen in others. These different songs do not have different functions, although each song does attract mates and repel rival by advertising territory. The songs are established in the bird's first year of life and then stay with it for the rest of its life. It seemed unlikely that birds would have a large repertoire of songs if one would be sufficient. And indeed, female song sparrows do prefer males who sing a variety of songs, as was proved in the laboratory in a series of experiments by William A. Searcy and Peter Marler.[4] These researchers recorded song sparrow songs and played them back to females who had been primed to respond to males by having been injected with estradiol, the female sex hormone. On hearing the male's songs the hormone-primed females responded to them by assuming the copulatory solicitation display (crouching with head and tail up and vibrating with their wings), and they responded (crouched) more if they heard the songs of their own mates relative to those of other local males (and they also preferred a *repertoire* of calls rather than the same number of the same songs).[5]

If a behavioral response is an adaptation resulting from evolution, then its advantage is measured in terms of more offspring, because that is why it came into being. So how and why did females have fewer offspring on average when they paired with a male who sang a given number of songs of *one* type, versus another who sang the same number of songs with equal vigor but of a different (though equivalent) type? Field studies by J. M. Reid and colleagues indicate that indeed there are more young produced per nest (provided the parents were first breeders) if the male had a large rather than small song repertoire.[6] Thus, it pays a female of this species to choose a male, or a territory occupied by such a male, with a large song repertoire, rather than one who has a small repertoire. The reason for the advantage (to leave more offspring) of the females' mate choice is, however, not number of songs as such, because singing doesn't cause fecundity. Instead, it turns out that song repertoire is a signal of something else; those males who sing several songs can also do some other things better than others. In short, a large song repertoire in these sparrows is a signal analogous to a vigorous flight display in snipe, number of nests built in wrens, or some other arbitrary male display that serves as a sign of health and vigor, affecting fecundity.

A large song repertoire as an "advertisement" of vigor that benefits the offspring still does not fully explain why males don't exaggerate and sing either more or more varied songs. The most likely explanation is that singers with low repertoire simply can't sing more songs. A high song repertoire, it turns out, requires a larger brain volume, and that brain volume (like antler growth in deer) can only be achieved under the advantage of optimal nutrition.[7] Furthermore, it costs energy to produce the neural machinery to sing, and singing itself takes time and costs energy. The song repertoire may thus correlate with the nutritional history of the bird, and that history is an indicator of being a successful survivor and a provider who benefits both the incubating female and the young who he must feed. Thus, the birds that have been able to devote a larger brain volume for storing song repertoires have also, on average, been better or more successful foragers. Those foraging skills then likely translate directly into being better food providers for their young and possibly other advantages—

better ability to digest and process food, and "better genes" coding for these behaviors, which are passed on to offspring.

Singing is a neurologically complex behavior, and nowhere is that complexity better shown than in the invention, memory, and execution of melodies and harmony, whether with tools or without. Sheer volume and vigor may suffice as a sexual attractant for some species, such as in the hammerhead bat. Males of this species have a large head with a bizarre face resulting from a mostly hollow head, huge lips, and an enormous larynx, all designed to produce and project a loud honking song that serves as a sexual attractant. Females fly along rows of hanging males who sing for her, and she chooses the one whose song she finds particularly irresistible.[8] I suspect birds are not alone in feeling an instant attraction to a good instrumentalist with a good voice and a good tune. I like listening to them; they attract my attention. We may not be impressed for long by anyone producing loud honks, or even a melodious but strictly stereotyped song. But our creativity and self-expression in song and dance are expressions of mental traits that serve as fitness indicators, and they may have been shaped by sexual selection.[9] Perhaps the birds' behavior hints at why we, like many of them, evolved to pay so much attention to music and to put so much effort into it.

The song repertoire of birds also functions in male-male competition.[10] Male song sparrows frequently reply to each other in vocal challenges, and switching songs apparently is a convention that indicates that the bird has a large repertoire. Michael D. Beecher and colleagues at the University of Washington tested counter-singing between neighbors by playing song sparrow songs from neighbors' territories, finding that subjects responded with shared songs.[11] The authors speculate that sharing repertoire with neighbors keeps the peace, maintains territory, and offers an opportunity to learn a neighbor's songs. A sparrow who has a large song repertoire can match any or many songs he hears, which lets the challenger know he had better back off, but the option is less available to one with a low repertoire. Thus when an intruding male does not back off but instead comes into a song sparrow's territory, possibly to conquer it, the owner may respond with a soft whisperlike song. Such muted song reliably predicts the bird's aggressive tendencies, and the opposing bird then faces a choice: fight or flight. How does he know whether the call is a bluff? It must have a cost in order to be an honest signal. One idea that has been suggested is that by calling softly the bird sacrifices his ability to ward off others, because they would not hear him or would presume he is at a distance. As a result a potential intruder, presuming there is no rival present, would invade. However, the soft caller identifies himself as being aggressive specifically and only to the one whose territory he is entering. He is saying, "I'm serious—I'm willing to fight you!" And ultimately there is only one reason a male fights: to achieve the status that accords breeding privileges. And song is just a part of the tests ultimately posed by the female to assess the male's ability to provide her something related to being a good parent to her kids.

*Courting.* Any way you measure it, little in life compares to the costs of parenting. Indeed, in many animals everything else may and often is sacrificed for it. By default the major burdens often fall on the female. Most female birds have to pay a particularly high price of parenting because their cost includes not only that of producing enormous eggs—frequently their clutch weighs more than their body mass—but also of preparing the place to hide, cradle, and protect the eggs. She alone may be stuck with incubating them and feeding and protecting the young. Any help that she gets should be welcome, and as already described, in most birds a mate is necessary to accomplish the task of bringing the young to adulthood. She must therefore choose a male who can deliver, and her assessment of him, which differs greatly among species, begins with the courting behavior that may ritualize or symbolize the commitment and capability of a potential mate and thus provide her with a filter for making the best choice. (In the polygynous species where the male offers the female no help, the courting symbolizes "good genes.") The relevant question may be not so much whether the female has a mate, but why she doesn't have several at once. I suspect she might gain by having more than one, but the males would then not be assured of parentage, so it would not be to their advantage to commit their full efforts to her young. As a consequence, the monogamous pairing is by default often the more evolutionarily stable strategy.

The first task in any mate search is to identify and isolate the relevant criteria, which as mentioned for polygamous species involves the currency of health, vigor, and genetic compatibility. For monogamous species, however, there is that but also more. Individuals matter, and bonding with them matters because their commitment and skills at the required task of prolonged parenting matter. All are evaluated through focused, though stereotyped, behavior patterns. Courting is conducted to locate a mate, but also to renew or strengthen bonds in already-mated pairs and perhaps to synchronize their reproductive behavior and physiology for the cooperative task ahead.

The primary limitation on whether or how many offspring a female can produce is often the amount of food that can be provided. Therefore, in this task a male's help is of critical importance. Not surprisingly, mate feeding is an almost universal courting ritual of birds, much like taking a date to a fancy restaurant. In those northern birds with huge clutches of eggs (titmice, for example, sometimes lay a dozen or more eggs per clutch totaling 140 percent of their body weight), feeding during courtship is also a necessity. She will produce an egg a day for a week or two and needs more than her normal quota of food intake to make it possible. Mate feeding may then reliably and directly provide a measure of the male's capability as a forager and provider for the young after they hatch, and it directly and indirectly increases the male's fecundity.

A female could simply take the food from the male's bill when he brings it to her. Instead, in courtship feeding, female passerines typically beg for food from their mates by mimicking the "baby bird" begging vocalizations. They open their mouths and flap their wings much as a baby bird might do. I suspect that this behavior tests the male's commitment to respond to the same signals of her imminent offspring. If he responds to her behavior, then he is likely willing and able to respond to her babies' similar behavior when they arrive. His courting display in response to her mimicry is an honest advertisement of his capability and likely future behavior.

Food provisioning capability may not always be tested directly. In some cases, typically in raptors, ability to be a provider may be displayed indirectly. Male eagles dive on the female from a great height in mock attacks, and when he reaches her she turns on her back. They lock talons and tumble. He is symbolically catching her as he might catch prey. Males who dive poorly and miss their mark are probably less likely to catch the prey that she and the family will later need. Ravens' ability to forage depends in large part on their willingness and wherewithal to fly long distances, since the carcasses on which a family's food supply depends may be separated by hundreds of miles. The raven's elegant flight maneuvers during courting rituals may thus be correlated with food-finding ability; the better and more willing flyers will not only be healthier and more vigorous than the lazy flyers. They will also be more likely to find the very widely distributed animal carcasses that constitute the young's primary food.

*Dance.* On this May 3, 2006, gray morning the phoebe that nests by our house every spring did not sing. It is probably having a hard time finding insects. However, earthworms should be emerging onto the wet ground now, as they do every spring after the first big rain when they disperse above ground. They should provide a food bonanza for the recently returned common snipes and woodcocks, and perhaps once both are "fueled up" with ample food, they will perform their vigorous nuptial displays.

The sky is overcast and drizzly when near noon I stop at my favorite meadow for snipes. Today they are indeed rocketing all over the place! They fly high, fast, and smooth, reminding me of jet planes, except that you can see their wings beating like those of hummingbirds. The whinnying sound of their wingbeats modulates the air flow over vibrating outer tail feathers and produces an unearthly flutelike sound—an impressive display that exudes power and energy. Up to five males are airborne at once, displaying over the expanse of cow pasture and fields with alder thickets and sedge hummocks at one end and a shallow watercourse running through it. They all seem to own the same air space and do not interact; unlike the dueling song sparrows, they do not fight. In the evening after sundown their close relatives, the woodcocks, perform their perhaps even more spectacular aerial dances.

The alders are starting to pop their leaf buds. Marsh marigold flowers shine bright golden-yellow, and their brilliant green leaves are poking up out of the mud and through last year's matted brown grass. Bluish-green shoots of the blue flag iris reach up, too. And somewhere, in some clump of sedge or grass, in contrast to the flamboyant males up in the sky above her, will be a snipe hen camouflaged in grays, brown, and sienna settled deeply into her nest cup after having laid her fourth and last egg of her clutch. Each egg will be camouflaged in dark olive greens and black, and it will be practically invisible in the shade of overhanging sedge leaves. Her belly feathers will be parted, and she will press her warm skin against her eggs as the feathers from her sides flare around and insulate them. Before she lays her eggs, does she evaluate the sky dances of the males above her to choose a mate? The woodcock female nearby, already incubating, is presumably oblivious to the snipe's dance because her male's dance is vastly different, though no less beautiful to me. It would not budge a female snipe in the least.

I dreamed one night that I was at a huge dance where everyone was performing wildly to the exciting music. I then recalled the dances I used to go to when I was a teenager and young adult. Everyone was gussied up in their coolest, most stylish duds. I recalled performing very vigorous, stereotyped movements at a lek called the "dance floor," and I was challenged to get my steps and movements just right. In fact, they had to be perfect. I did not for even one moment ask myself why I and everyone else of my general age were doing this. If anyone had asked me I would have thought it a stupid question. I did it because it is fun. Duh! That fact was so obvious that it didn't need thought.

Now, decades later, I thought of my youthful exuberance, expressed in those very stereotyped steps, movements, and rhythms on the dance floor, and it seemed to me they uncannily mimicked those of cranes. Crane courtship movements are literally dances. I had just visited my friend and colleague George Happ and his wife Christy Yuncker near Fairbanks, Alaska, where they had for fourteen seasons followed the nesting behavior of a pair of sandhill cranes, Millie and Roy, who return every spring to the bog and marsh adjacent to their house. George and Christy have ringside seats for watching and studying the cranes' antics. I had planned my visit so I could see the arrival —the first very touchdown—of the pair after their migration thousands of miles back home from Texas. The pair arrived on the evening of April 28 while we were having supper. We saw the pair gently touch down on the ice of the still-frozen pond. Roy almost immediately commenced excited bugling, and then I saw him dance. He commenced to crouch, leap high, and flap his wings and gyrate in a show of strength and power. It would be an anthropomorphic mistake, however, to suppose the cranes dance only for joy.

A little over two months later, on July 2, when their two colts were almost three weeks of age, one died in the night. The following day, as George and Christie report: "All three cranes danced with fierce intensity—not frenetically, but controlled, fluid, graceful and balanced." It was unprecedented for Roy and Millie to dance together in July, so the dancing probably had some connection with the dead colt to which they repeatedly returned and examined. They clearly recognized it as something special, and even engaged in displacement activity near it, such as going through some motions reminiscent of nest building.

Birds gauge their partners on the basis of compatibility, and a large part of the compatibility relates to whether or not they successfully rear young. Therefore, the loss of the nest or young could be grounds for divorce. Inasmuch as both crane parents are intensely involved in foraging for and feeding and protecting their young, and since the dancing has functional significance in solidifying the pair bonds, did this dancing reaffirm what might otherwise have been weakened bonds?

The steps of the cranes' dance performance, like any dance of ours, are set by convention and are probably honed to perfection by maturation and by practice. If there were no strict convention, there could be no perfection. But every species' convention is different, and they vary from the crude performance of a cowbird's fluff, crouch, and bow, during which it makes a creaky squawk, to the mind-bogglingly elaborate performances of some birds of paradise. They are brilliantly portrayed on a video produced by the British naturalist and film producer David Attenborough. These birds are the Bolshoi Ballet of dance, performing with breathtaking jumps, steps, and gyrations while decked out in the most spectacular of plumes. They must be seen to be believed. And they are well appreciated, attracting an audience of females. These males are under pressure— they have to do it precisely right. And they get it so right that we can't even distinguish the adequate from the poor performance, because the females are so picky that only one male may get all the matings. The rewards for good performance are high and yield immediate gratification. I suspect the bar of dress and performance is raised so high that it explains why the males don't even bother to put on their "gown" of gaudy feathers in order to perform until they are six years old. They needn't try until they have a chance, and they have little chance of scoring until they have matured and maybe practiced for years as well.

As in the male displays of many species, the female birds of paradise prefer to mate with males that produce courtship displays showing high intensity and fast movements. As I recall from my own dancing attempts, I tended to overdo it and lose coordination, startling partners and bystanders alike. The same phenomenon occurs in some birds. In a recent study by Gail Patricelli, Seth Coleman, and Gerald Borgia, an attempt was made to identify the males' moves that turn on satin bowerbird females.[12] The researchers constructed robotic female bowerbirds by draping skins of these relatives of birds of paradise over servos that could be moved by remote control. The robotic females then "startled" during the males' displays. They discovered that the displaying males are attentive to the females' responses; when the male's too-vigorous display startled a female (the robot), the males decreased the intensity of their next performance. (It is worth noting that the males are relatively not choosy at all; compared to the females they are practically promiscuous.)

I recently watched the movie *Grease* to refresh my memory of a certain youthful time, and I was mesmerized by the dancer's movements and song. Who could not be impressed by the performance of the young John Travolta? It made me think of the woodcocks and the snipes, of birds of paradise, of dancing cranes, and of the dances of male craneflies I had recently watched. The wood frogs in their spring chorus came to mind. And then I suddenly knew why I had enjoyed our communal gatherings at the lek of the Weld, Maine town hall on those warm summer Friday nights in June. Yes, it had all been fun. It was so much fun that I literally had to do it, which probably meant it was something I felt instinctively, driven in the same way that a woodcock, a crane, or a bird of paradise dances. Now I know: it was all about the girls, and I didn't even know it then.

Song sparrows are sophisticated musical performers; much more innovative than woodcock and snipe. Maybe we have also been selected to be able to perform and remember long and varied repertoires and to keep up with changing ones to signify that we are current. Maybe such expertise signifies something more, something to do with behavioral flexibility that results in sexual selection. A one-hit wonder is less likely to get the girl (or girls) than a group with an endlessly varied repertoire like the Beatles. We are socially "monogamous," but if recent DNA technology can be trusted, and I think it can, then one in ten children does not have the biological father that is socially assumed; we are much like birds, though even more flexible. Maybe a million or more years ago, or maybe even more recently, some "Travolta" or "Sinatra" had more than one wife, or maybe even had, here and there, some of another's children. Why else would the powerful human emotion of jealousy (promoting mate guarding, in the same way that love promotes mating) evolve? Why else would we have a strong social taboo against mating outside the pair bond if there had not been a reason to fear it?

The performances of birds that signify "good genes" often translate to direct benefits. Males, especially in monogamous species, are indispensable to females in child rearing. And there are many ways that males can help directly in the nesting process, starting with nest or home building itself.

*Display nests.* Although nest building is most often the responsibility of the female, it is one of the tasks in which a worthy male may make himself useful because males should presumably be able to do everything a female can do. Indeed, in some species, such as wrens, the penduline tit, and in many weaver birds, the male builds a partial nest, and this nest becomes either a platform for the male courting display, as in the weaver birds, or a mate attractant. The nest serves as a record of the male's industry, skill, and competence, and it is thus also an indicator of "good genes." In contrast to some other means of attracting mates, it provides actual help as well. A female inspecting nests compares males indirectly. She chooses for herself a resource she needs for reproduction and also indirectly assesses the male's potential vigor and industry. Perhaps not surprisingly, where male-built nests have become sexual attractants, they may be impressive works indeed, such as those of weaver birds and the penduline tit.

A titmouse, the penduline tit, builds a very conspicuous nest, one that like many weaver birds' is relatively inaccessible to a number of predators because it is attached to the tip of a slender hanging twig, often over water. That is, it need not be hidden from view. This nest is truly a work of art in the bird world. In most tits the nest is built mostly or exclusively by the female, but the penduline tit's nest is initiated by a male. He builds a pear-shaped structure of long plant fibers, and if it attracts and suits a female, she then incorporates fuzzy plant material (such as the down of ripened willow fruit) and mixes them with saliva. By constant picking she makes the nest's walls into the texture of felt. She also adds a long, tubelike entrance to the basic nest structure, the mitt of felt. A robin female can build her nest alone in two days, but it takes the tits between twelve and twenty days to build their elaborate nest. The female starts to lay her pure white eggs before it is finished, and as soon as she starts laying, the male then leaves her and starts building a second nest to attract a second female. He may do so up to five times in a season. Unlike in other tits, in which both members of a pair feed the young, the male penduline tit has no time for that—he builds nests. The female is assisted instead by her offspring from her previous brood.[13]

Perhaps the ultimate in the trajectory of nests serving in sexual selection is the example of the Australasian bowerbirds. These build what I suspect are in fact pseudo-nests which started out as nests early in evolution but that no longer serve as such. They now function only as display props to attract females. In a real nest there are always numerous constraints that set limits to display. And of course the biggest constraint of a real nest, one with eggs and young, is that it should be hidden. But of course the male bowerbird constructions are precisely the opposite. They must be conspicuous, and they can be because they don't contain his babies. In the nuptial displays of one group of flies, the males offer "prey" bound by silk to females. In these the packaging has also become everything. The female flies evolved to use the silk packaging as a signal that indicated food. It then became a huge silk balloon containing no food at all, much like gifts we give in which the gift wrapping has become more important than the contents. I suspect that the huge decorated bowers of bowerbirds are similarly derived from display nests that became packages for show only. As such this has released these "nests" from the necessity of being hidden from predators, opening the way to any of a great variety of conspicuous shapes and sizes and allowing them to be built in conspicuous places with noncamouflaging materials. These bowers are often decorated with such brightly colored objects as berries, flowers, and shells to attract females for mating.[14] It is conceivable that in a tropical environment, with no time constraints of the nesting season, males need never finish a bower. They can build on it year-round. Females can attend males' pseudo-nests and then also build their own, functional and hidden.

As long as the nest built by a male serves a dual purpose—sexual attractant and shelter for the eggs (as in wrens and weaver birds) —there will be constraints on the conspicuousness of external nesting material. However, such constraints might not apply to nest linings, and perhaps here showy and exotic materials could serve as attractants. If those materials serve a practical function as well, then the selective pressure to evolve the behavior of incorporating them into the nests should be all the greater.

*Costume.* One of the most well-recognized and amazing features of birds is their plumage. Unlike most animals they have many garbs. Some have a highly insulating and beautifully camouflaging plumage when they hatch out of the eggs. This is soon dropped to be replaced by a juvenile garb, which in turn is exchanged to a gender-specific adult garb. Birds may acquire their adult garb within months or much, much later—perhaps up to six years later. Furthermore, the garb can change appearance by instant rearrangements of feather and body posture. Obviously appearance is highly important. We tend to think that the bright male nuptial garb is for show, both to other males and to females, and is thus directly or indirectly related to nesting. Although the specific colors may be arbitrary, it is important that, in contrast to previous garb, they distinguish themselves from the background. This may include being distinct from other species living in the same area, possibly to avoid hybridization. For example, the black throat patch and eye mask of golden-winged warbler males were lightened by researchers (with bleach), and these altered males lost their breeding territories and failed to get mates.[15] Apparently their prominent black facial pattern reduces or prevents the hybridization that sometimes occurs in this species with the blue-winged warbler in the wild.

Distinctiveness of the nuptial garb can be achieved not only by song or dance but also by colored accoutrements (usually feathers, but also bill, wattles, eye rings and eye color, feet and leg coloration, and bare, brightly colored skin). The rose-breasted grosbeaks that breed near my house and frequent my birdfeeder are one example. The male has a black head and a striking pink breast, a pure white belly, and a pearly white beak. The closely related black-headed grosbeak has a cinnamon-orange breast, belly, and neck, but its song is very similar. The female's garb is almost identically brownish and sparrowlike in both species. The first-year plumage of the young of both species is also almost identical and similar to the female garb. That is what the birder in the field would see, and that is what the field guides show. But there is more.

In the summer of 2008 a pair of rose-breasted grosbeaks built their nest of twigs and rootlets in bushes near our house as usual. The female laid a clutch of five greenish-blue cinnamon-blotched eggs and successfully raised their young. The family stayed near and frequently visited our bird feeder, well supplied with black sunflower seeds. Unfortunately one of the recently fledged grosbeaks banged into a window.[16] Windows are invisible to birds and bird–window collisions are estimated to be one of the greatest human-associated causes of bird mortality next to house cats and habitat destruction. When I picked up the stunned bird, I was surprised to note that the seemingly drab plumage was not drab at all; the feathers on the underside of the wing were bright rose-pink. I photographed the bird before it recovered and I let it go. A couple of weeks later a second of the young hit the window and was also stunned. This one, instead of rose-pink under the wing, had bright lemon-yellow. The colors probably identified the genders (pink for male, yellow for female), but there is presumably no need for young birds either to be showy or to identify their gender.

After I had photographed the second young bird and released her, she flew weakly and landed on a big rock. Almost instantly another grosbeak appeared and landed near her. It seemed interested, checking her out. I presumed at first that it was the stunned bird's mother. But then I noticed the distinctive bright pink breast of a first-year male. This was a young bird from the previous year that had molted once. It would molt into fully adult plumage with a large pink front the next year. I doubt that five color morphs (three male, two female) are necessary in one species since all but one (the adult male) of these color morphs are nearly identical until examined up close.

In many species the males' nuptial displays and the brightly colored feathers that serve as sexual attractants are proximally controlled by testosterone. For example, in the red-backed fairy wren from Australia, three distinct color morphs, one bright red and black, the other two dull brown, are associated with decreasing amounts of testosterone both during the molt and during breeding.[17] The bright red and black males sire the most young, including the most extra-pair young, whereas the dull-colored helpers at the nest sire the fewest, if any. Since testosterone levels depend on physical condition (including nutrition and absence of parasites), the hormone thus serves as the mediator between physical condition and the honest advertisement of gender.

Given that nuptial garb advertises gender and male quality, it is not surprising that one of the first things that birds do even before they reach the northern breeding grounds is to acquire their "nuptial garb." The males may or may not be outfitted in bright feathers, but almost all the females remain drab by comparison.[18]

The males of most finches, warblers, icterids, ducks, and pheasants are clad in bright conspicuous plumage in contrast to the modest garb of their females. Their bright garb is an important part of the ornamentation that functions as a sexual attractant for females, along with the long or showy tails and wings of some birds and the combs of roosters. In many finches the bright colors—mainly reds and yellows—are derived from plant carotenoids.[19] In 2005 the normally yellow-orange northern orioles started turning up bright red during their fall migration. Their feathers contained rhodoxanthin, a deep red pigment found in honeysuckle berries, of which there were bumper crops that year.[20] Berries are colored with carotenoids presumably for the advantage of plants, so that birds will more easily find them and disperse their seeds. It turns out, though, that the colors also help the bird's "seed" in sexual reproduction.

Ever since Charles Darwin explained the sometimes extravagant colors and displays of birds in terms of sexual selection, a large body of literature has accumulated on the topic that shows how the plant pigment is involved in sexual selection. Although it was long assumed that bright coloration in males functions in sexual attraction, it was never quite understood why or how it worked. The theoretical problem was that color, as such, seemed too cheap to be credible. That is, if all could wear the same sexy dress, then what advantage could it constitute? However, in 1982 the late William D. Hamilton and Marlene Zuk published a hypothesis that gave the topic new life. They pointed out that bright coloration may not be cheap at all. Not only is there variation in the coloring trait, but parasites reduce the capacity of individuals to show their colors. A female can assess parasite load (and presumably susceptibility to parasitization) of potential mates and thus more easily choose a healthy mate on the basis of his color and ornaments.[21]

Subsequent research supported the Hamilton-Zuk hypothesis. Jungle fowl infected with roundworms have reduced comb size; barn swallows infected with blood-sucking mites have shorter tail ornaments; green finches are less brightly colored when infected with blood parasites; male house finches with blood parasites are less red than uninfected individuals.[22] Indeed, it turns out that bright color due to carotenoids in the diet correlates not only with reduced parasite load (and possibly good immune function) but also with increased over-wintering survival and general good nutritional condition. The color brightness is thus an honest advertisement of quality and not just cheap show; there is a functional advantage for females to prefer healthy mates who will not only be more likely to live to help her produce offspring but who will also be more likely to carry genes that favorably affect her offspring indirectly.

Not just any carotenoid will do for coloring. In house finches, a very specific carotenoid, beta-cryptoxanthin, is required, and there is also a cost associated with foraging for it or a cost for the metabolic transformation to produce it from the dietary precursors.[23]

Although many birds acquire color through diet, there are many examples of bright plumage unaffected by specific diet. For instance, some of the most brilliant colors of all, such as those of hummingbirds, some ducks and trogons, peacocks, and birds of paradise, are achieved without specific pigmentation. They are "interference colors" such as those we see when a droplet of oil spreads on water. It may be that in these cases, colors function predominantly to draw attention to displays that the males make, flashily advertising physical vigor as vocalizations do.

Birds' eyes contain not only the same color receptors that we have, but also another type sensitive to the ultraviolet, hence they should see more color than we do. Their attire, when bright so as to contrast with the usual green and brown environment, is likely meant for their eyes only. On the other hand, colors and color patterns that blend in with the environment are likely meant for camouflage from predators. To be noticeable must be costly. But through molts the period of such flashy if risky attire can be reduced to a season. At the extreme in showiness are the polygynous birds—the birds of paradise of New Guinea and African whydahs. And it is the extremes that are the most instructive. These males, who do not help at the nest, are advertising "good genes." Nothing else. The same signals that attract the females can easily be intercepted by predators. That is, their show compromises their survival and is thus another honest advertisement of quality. However, survival as such is not the relevant currency of natural selection. It is instead the number of surviving offspring sired over a lifetime. A grouse that drums may not survive as long as one who does not, but if the females are attracted only to drummers, then although nondrummers may live much longer, they still leave no offspring. Natural selection will then preferentially produce drummers despite their survival handicap.

One theory that has received much attention, first promoted in 1975 by Amotz Zahavi, is called the "handicap principle."[24] Since in order to be honest a signal such as singing or displaying to attract mates must be costly to the signaler, birds display a "handicap" when they court, which then serves as an honest advertisement of their quality. They are presumed to be wearing a handicap like a badge that says they have superior qualities relative to the competition because they can function well despite the handicap. However, animals notoriously avoid advertising handicaps. If possible, they instead "cheat" and exaggerate strengths while hiding handicaps because the choosing sex does not look for handicaps as such. A peacock, because of its long and massive as well as conspicuous tail, is indeed handicapped in predator avoidance, and a bird who sings loudly is also easily located by predators. However, these effects are costs of the mating game. Perhaps focusing on "handicap," the consequence, rather than the desired trait, muddies the waters rather than clarifying them. The handicaps that we can deduce in a peacock's tail are not seen as handicaps by those for whom the conspicuous signals are meant. The "intent" is to show off—because he can and the other guy can't.

Although bright colors can signify "good genes" and serve as a sexual attractant, in male tanagers, many finches, orioles, ducks, and most warblers, for example, many other species dress less ostentatiously (woodpeckers, many sparrows, flycatchers and larks, crows, wrens, vireos, swallows, chickadees, nuthatches, most thrushes, larks, brown creepers, flycatchers, and many others). In these species the male and female garb are almost identical and do not change in and out of the nesting season. In still other species there is modest sexual dimorphism in plumage but it lasts year-round; it is not necessarily restricted to the breeding season. Many finches, including crossbills, woodpeckers, and kinglets, fall into this category. Molting in most species occurs once per year at a specific time. Feathers drop out and are replaced on a schedule particular to each species. For example, the flight feathers are shed sequentially so that there are always enough for flight (except in some waterfowl who shed them all at once when they have young and need to accompany them on the ground). Birds such as goldfinches and others who have bright nuptial garb shed their drab winter feathers in the spring and then shed the nuptial garb in the fall to get back into their nonbreeding plumage. Could the double molt that is required specifically for the nuptial colors also be a primary cost that signals their worth through "honest advertisement"? On the other hand, if there are great advantages to bright nuptial colors, one might expect that all species would adopt them. Since many don't, one may legitimately wonder if there is a value to the unobtrusive unigarb worn by both sexes in many species beyond the obvious one of camouflage from predators?

If strutting one's stuff is a means of competing, then in social or semi-social species disguising one's gender may help promote harmony. Indeed, deer males cast their antlers, or females grow smaller antlers (in caribou), which helps the younger males blend into herds containing adults of both sexes. That is, sexual dimorphism would facilitate strife, such as would occur among males who fight among themselves for harems. On the other hand, peer competition would promote sexual dimorphism, and it would act to evict adolescent males from the herd. When the two sexes look similar, there are no badges of gender, and that restricts aggression by breeding males to breeding times. Peace then reigns and the advantages of the herd, such as safety in numbers, are realized. Could these same ideas, articulated by the mammalogist Richard Estes, also apply to birds?[25] The quickest, surest way may be to first examine the extremes.

The extremes in sexual dimorphism probably occur in the birds of paradise, vidua finches, pheasants, and some ducks. In all of these there is intense male-male competition for females, and the birds maintain their showy garb year-round. They are solitary both during and out of the breeding season. There is a delay—sometimes many years—before the males acquire their "horns." They do not flock. At the other extreme are birds with uniform, year-round garb in both sexes. These include some finches such as pine siskins (who may nest at any time), waxwings, swallows, swifts, geese of various species, gulls, pigeons and doves, cranes, penguins, flamingos, grackles, crows, jays, and magpies. All of these may flock at any time of the year, and they reveal their gender facultatively and as appropriate by behavior and vocalization. Estes's hypothesis for herd mammals seems to hold for flocking birds.

Many interesting exceptions are also compatible with these ideas about plumage. For example, red-winged blackbird males hide their crimson wing badges by covering them with other feathers when they are in a nonbreeding flock of other males. That is, the males maintain their distinctive capacity to show their red patches all year, but partially hide them when they fly in flocks, which are segregated from the nondescript females. Goldfinches and some other finches and icterids, in contrast, wear unisex garb in winter, but the males sport distinct, showy plumage in the breeding season, at which time they separate from the flock. In order to acquire their gender badges they must molt twice a year (once into the garb, and then again out of it). Most other birds can get by with a less costly wardrobe by exchanging their garb with only one molt per year.

I suspect an in-depth study would reveal some similarities even with our own dress codes. The most garish and expensive accoutrements are worn by rock stars and other celebrities. The rest of us, particularly men and women who work in organizations where we are virtually a flock, wear similar suits and ties. The unspoken and perhaps unconscious objective is not to accentuate either femininity or masculinity. Go out on the town, and our individuality comes to the fore, which requires flashy duds showing distinct sexual dimorphism.

I find it curious that at one time there was a fashion of men wearing a "tailcoat," sometimes also referred to as a "swallow-tail," on formal occasions. This strictly male garb has been worn since the 1850s, looking not unlike the long tails of male swallows and widowfinches (whydahs). In the laboratory, experiments with clipping these birds' tails and gluing extensions onto others with superglue to lengthen them proved that the tail ornaments of these two very different birds enhance sexual attraction by females.[26] I know of no analogous experiments with human males. But male Australian zebra finches equipped with artificial white head plumes were more attractive to females of that species than control males without these headdresses.[27] In these finches the males have short tails but bright red beaks (as opposed to the females' yellow-orange ones), and the females also prefer to mate with red-billed males. Both sexes prefer some traits that don't exist outside of a laboratory experiment, suggesting an aesthetic sense; both partners preferred to mate with partners having specifically colored leg rings.[28] For example, males with red leg rings had more extra-pair copulations (with accompanying extra-pair fertilizations). A further study with collared flycatchers showed a similar boost in sexual attraction for a novel trait.[29] In this case it was a red stripe painted onto the white patch on the male's forehead. This attraction to the novel trait, which simulated a mutation, was learned. Females that had associated with males whose normal white face patch had the red stripe later on preferred to mate with such similarly attired males, whereas they had a slightly lower preference for them if they had not learned about the red stripe. The authors of that study suggest that learning may play an important role in the evolution of sexual selection of divergent traits. I wonder if a similar phenomenon may account for human changes of fashion. I recall that in high school and college there was rather intense social pressure to be "cool." Coolness was expressed at least in part in our attire; it was to tread the fine line to be a conformist and to be novel at the same time. The result was, and is, that there are changing trends that result from learning, involving mimicry in which novel "mutations" are selected after the mimicked modes have been mimicked to death. What is now current on my university campus as I write this chapter will have changed by the time this book is published.

*What good are males?* Sexual reproduction has long been touted as one of the great unsolved biological puzzles, and it has usually been reduced to a single question: Why are males necessary? I have so far emphasized that raising young often requires partnerships. It requires help. But it is often pointed out that an asexual female, or a pair of them, who could produce two female offspring per individual in a lifetime would produce four female granddaughters. On the other hand, if she were sexual and produced both males and females, she would on average have only half as many female grandkids. (Only daughters have eggs. Sons can fertilize eggs, but they cannot bear young.) By this logic, the asexual line should quickly outcompete the sexual. The adaptive significance as usually given is that although sexual reproduction is proximally inefficient, it has long-term benefits because it produces genetic variety. Genetic variety is better suited to respond to a changing environment. It is putting one's eggs into many baskets.

The problem with that argument is that selection as we know it favors what is immediately beneficial to the individual. It is blind to the future. However, recent research is providing a convincing "rescue" from the conundrum of why we have sex by providing an immediate selective advantage for it, and hence it helps to explain fundamentals of sexual selection that Darwin could not even have suspected. What rescues males from being wasted appendages and makes them critical players in proximate survival is to be found in molecular biology.

It is becoming ever more apparent that the environment is a far more variable, complex, and critical factor than previously supposed.

As posited by the late evolutionary biologist William D. Hamilton, we probably have males (and sex) because of diseases and parasites.[30] Hamilton modeled population growth of asexual versus sexual reproduction, and the asexual won out. Then he introduced disease, and it reversed. He concluded: "The essence of sex is that it stores genes that are currently bad but have promise for reuse. It continually tries them in combination, waiting for the time when the focus of disadvantage has moved elsewhere."

Viruses, bacteria, and other disease organisms have very short generation times. They evolve so quickly that constant preparedness is required to keep up with them, and new genetic defenses are needed that evolution, in organisms that have life spans of years, cannot keep up with. A bird that has a generation time measured in years cannot win an evolutionary arms race with a bacterium or virus with a generation time measured in minutes to hours. Disease organisms change so rapidly that "it takes all the running you can do, to keep in the same place," as the Red Queen said to Alice in Wonderland.[31] New combinations of genes are required not just for the future but in every generation, and this puts another variable into the parenting (and hence mate-choice) equation.

The study of sexual selection, once focused on exterior ornaments and displays, has recently shifted to consider the genetic underpinnings of the immune system as a possible solution to deal with the unpredictable scenario of an ever-shifting field of parasites. In one solution to generate genetic variety to deal with rapid parasite evolution, the "best of a bad job" scenario, the emphasis is simply to generate or maintain genetic diversity. A particular pathogen may kill one of your sexual offspring, but due to the genetic diversity of your offspring is unlikely to kill them all. Evidence is mounting that many vertebrate animals do considerably better than that, and it is accomplished by a mate choice geared to gaining specific genetic compatibility for maximizing immune capability.

The best-case scenario (for a vertebrate animal) is if the

choosing sex "knows" his or her own immune genes and recognizes whether those of potential mates confer resistance to the current parasites. It should then preferentially choose mates whose genes complement its genes. The offspring will then inherit the best possible defense. As over-demanding and fantastic as such a scenario may seem, it is more than a theoretical fantasy. Compelling evidence in mammals (including humans) and fish is accumulating showing that immune alleles can be identified by smell, and mate preferences may then be allocated accordingly.[32] Nothing is known so far whether or not any birds may similarly employ smell in mate choice. However, costly phenotypes such as vigorous and complicated songs, displays, and vibrant plumage are indeed signs of health, and they are thus a reflection of current fitness with respect to current pathogens.[33] The choosing sex thus has a "window" into the state of a potential mate's immune genes, and in some birds a link has been found between such displays and immunity alleles.[34] These "good genes" that the choosing sex acquires may then be combined optimally with its own good genes to achieve best complementation.[35]

As always, genetics ultimately serves as a basis for mate choice, and the problem birds face is how to evaluate the specific, relevant characteristics that might help them choose, enhancing their and their progeny's reproductive success. The vast and dazzling diversity of mate-choice criteria—offerings and ornamentation, displays of song and dance, costumes and plumes—are precisely those that we perceive as beauty to the eye, ear, and mind. They document the senses and sensibilities and behaviors we share with birds.[36] But there are undoubtedly still other perhaps subtle behaviors and features birds find relevant, but of which we are not yet aware.

## 7

### NEST SITE
### AND SAFETY

IT IS SNOWING AND SLEETING THIS APRIL 4, 2008, and the beaver bog is still frozen under thick ice, but daffodils and snowdrops are shooting up under the snow, and the first song sparrow sings by our house. Two pairs of Canada geese came today and walked all over the ice, and one pair gave chase to a fifth one that was alone. The usual common grackles and red-winged blackbirds are now also back. But today brought special visitors: two pairs of European starlings. Within minutes of arrival they were flying from one nest box to the next, and one pair was soon paying particular attention to one that I had made from an already hollowed-out cedar log with an irregular oblong knothole for an entrance, which I had hung onto a post at the edge of our garden.

*April 24, 1998.* I treat myself to a jog in the afternoon sunshine down our dirt road through the woods. The poplars have just flushed out in delicate yellow-green leaves, and the serviceberry trees show up as white puffs against the woods as they are clothed in massed blossoms. The sugar maples are in full-tasseled yellow although as yet not showing a hint of leaves, while the red maples are already shedding their red flowers and will now start to leaf out. Robins have been back for three weeks and are now singing at full throttle. I found five fresh nests along a two-mile stretch of road. One had just finished the mud cup, three had lined their nests with grass, and the fifth had its first egg. The pure light blue on the tawny brown dry grass always catches my breath. I expect that most of these highly visible nests will be robbed, but those built on our and the neighbors' wreaths hung on the garage doors always fledge young, usually in two or three successive broods from the same nest, although I have never seen a nest in the wild used twice.

*May 16, 2003.* A crow came by twice this morning, and each time one of the grackles (of the four nests in a group in the nearby bog) rose up to intercept the intruder and handily routed it. Is that why it pays to nest near each other: protection? I saw a great blue heron, holding a frog in its bill, land about six meters from the goose nest. The gander had remained mute and out of sight during the last twenty-six days that the goose has been incubating, but now suddenly he flew in and stopped near the heron, just watching it. We seldom see a heron here, so the gander was probably just playing it safe and checking it out. However, he didn't watch long. Even before the heron had swallowed his frog a grackle dove at the bird, and with heavy wingbeats the heron fled. The grackle stayed in close pursuit and repeatedly dived at it. The agitated heron made croaking sounds and at the same time dipped in flight to avert the grackle's dives. No contest. This David handily routed a Goliath. A week ago an osprey was similarly routed, first by a grackle and then by a red-winged blackbird. A day later, at dawn, I saw a robin nipping a fleeing grackle. (I checked the robin's nest—it had just been emptied and now contained only some fresh broken eggshells.) Cause wins over size.

*May 21, 2008.* Only one of the starling pairs stayed, and they now they have four feathered-out young in their box. The tree swallows in a nearby box have finished laying their clutch of four eggs and started incubating today. Many of the other birds have not been so lucky in their nesting. The raven pair on the nearby ledge had chosen a spot that seemed inaccessible, but it was raided nevertheless shortly after the young hatched. Specific nest sites, such as tree holes and cliffs, are often in short supply, and they determine where and whether a bird can nest. Ground nesters can nest almost anywhere, provided they have a suitable territory. But they must contend with many more predators, including snakes, weasels, squirrels, mice, raccoons, and coyotes.

The Canada geese of this bog are limited by specific safe nest sites, which in this area are usually beaver or muskrat lodges surrounded by water. Several pairs of geese fought long and hard for "ownership" of the bog, and only two remained. One pair attempted to nest in contested territory of the other, and their nest with eggs was destroyed. It is too late now for them to re-nest this year. The remaining pair, Jane and her gander, have nested here since 2001, and this spring she built her nest on the long, sinuous beaver dam that crosses our entire bog. It is an ancient dam that is overgrown with grass in the summer, and the beavers keep putting on fresh mud and sticks. But the pair had chosen a dangerous spot. Coyotes and raccoons could cross along that dam. My fears for this nest were realized when I did not see Jane on the nest yesterday (viewing with my binoculars from our living room couch). I went down to check and found her five eggs scattered. Some of them were broken to reveal half-grown embryos. Next to the nest were also a half-dozen wing feathers laying together as if they had been ripped out in one bite. I followed a trail of contour feathers that led along the dam and then through the cattails and finally ended in the bushes at a huge pile of down feathers, her half-eaten body, and her severed head.

The third pair of geese near here had settled in a neighboring beaver pond a half-mile down the road. They had built their nest on the very tip top of a new lodge. I had high hopes for this nest. But although the goose was on the nest several days earlier, she was not visible today, and so I waded out to the lodge and climbed onto it to check. The nest was gone, and there were freshly beaver-gnawed sticks on the top of the lodge where it had been. I removed a few of these sticks and discovered that the beavers had chewed up from below, in an apparent attempt to enlarge their chamber. The goose nest and eggs must have fallen down into it, and then the beavers tried to patch the ceiling by putting fresh sticks on top of the lodge where the nest had been. As I peered down into the beavers' nest chamber I didn't see goose eggs, but instead noticed several tiny beaver kits. There is nothing cuter than a little brown fuzzy beaver kit. Luckily reason prevailed and I thought the better of taking one home with me, just as one after another of them scrambled off their nest platform and submerged into the water. Within a minute they started popping up like corks onto the pond surface all around me, and then they swam to rest at the pond edge. Meanwhile, their parents milled about, slapping their tails onto the water. Although the geese's nest attempt failed that year, they were back the next year and nested on another nearby beaver lodge. This time they successfully hatched out their young, but the family was intercepted on their migration through the woods to their feeding area by the neighbor's dog, who killed them all.

*June 9, 2007.* Many of the grackles and red-winged blackbirds have nestlings now, although others are still just finishing building nests. I suspect these are re-nestings of previously failed attempts. These birds may find relatively safe nesting in the dense thickets of the bog over the water, because in the woods there are weasels, raccoons, squirrels, deer mice, blue jays, crows, hawks, and even deer that will eat them. But safety is relative. In the open vistas of the marsh, the grackles and red-winged blackbirds can see a crow or a raven approach, and they cooperate in mutual defense; if one comes near they instantly go after it in droves, diving onto it and repeatedly hitting it. This morning they encountered something potentially more dangerous and less easily routed than a crow or raven. It was out of sight in the dense sedges, and a noisy crowd of grackles and blackbirds gathered. It was difficult to count, but there were at least twenty-five birds in the milling mob and I suspect it was most of the bog's population. The noisily scolding group was slowly moving across the bog, directing their attention to something they were following near ground or water level. I had in previous years seen such mobs, and nests were emptied of eggs and young afterward. I wanted to see the perpetrator this time, so I ran down to the bog and launched out into it in my kayak. I immediately saw piles of fresh mammal scat on a muskrat lodge. It contained fish scales and small bones. I heard mewing and chittering sounds and then saw wavelets among the cattails, and momentarily a black head—most likely an otter—appears at water level.

*July 23, 2005.* It poured yesterday, but now a gorgeous morning with clear skies unfolds. House wrens are busy at their nest box by our window, carrying feathers to line the nest for their second brood. Alder flycatchers and mourning doves are singing again, also getting ready for their second broods. Although many first broods were destroyed this year because of the rains, our first brood of phoebes was raided and eaten by chipmunks. The pair built a new nest in a more hidden, safer location that I provided for them—our woodshed.

In his book *The Natural History and Antiquities of Selborne* (1789), the naturalist Gilbert White (1720–1823) remarked on the nesting habits of jackdaws, which he called "daws." In his journal entry of November 28, 1786 he wrote: "And here will be the properest place to mention, while I think of it, an anecdote which was told me by a gentleman, that, in a warren joining to his outlet, many daws build every year in the rabbit-burrows under ground. The way he and his brothers used to take their nests, while they were boys, was by listening at the mouth of the holes; and if they heard the young ones cry, they twisted the nest out with a forked stick. Some water-fowl (viz. the puffins) breed, I know, in that manner; but I should never have suspected the daws of building in holes on the flat ground." And then he continued: "Another very unlikely spot is made use of by daws as a place to breed in, and that is Stonehenge. These birds deposit their nests in the interstices between the upright and the impost stones of that amazing work of antiquity: which circumstance alone speaks the prodigious height of the upright stones, that they should be tall enough to secure these nests from annoyance of shepherd-boys, who are always idling round that place."

Jackdaws are small crows with a white iris and a grey nape, and they are now almost if not totally dependent on homes provided by man (and other animals) for their nesting sites. In Europe they have traditionally used church steeples, old ruins, and almost any poorly repaired buildings that they could enter through holes. In Israel they established a colony in another site of human antiquity, the caves near the Kibbutz of Beit-Guvrin, which is named after the Jewish-Roman city by that name. It dates back to the Second Temple Period (spanning about six centuries before Christ during the reign of Cyrus the Great of Persia and the time of Christ), when limestone was mined there for house construction. The mining left caves that were at one time used by Jews escaping from Roman imperialism, and now they have become the home of jackdaws. They nest by the hundreds inside the caves' nooks and crannies. In the last decade these jackdaws have spread and they now nest in many other areas, including the city of Jerusalem. Sometimes they nest in chimneys. And I once saw a pair nesting in a former nest hole that a black woodpecker had excavated in a great beech tree in the forest. However, now that "progress" has overtaken many areas of western Europe these cheery little crows are being again pushed to find new nesting places, and in some places they now nest in the open tree nests that they take over from other crows, such as rooks.

Birds of the crow family, the corvids, are in general innovative and flexible in matters relating to nest safety. Rooks find safety by nesting in colonies at great elevation in trees, ravens build high in inaccessible cliffs as well as in tall trees. Magpies build an almost impenetrable roof over their nest out of solid sticks.

The nestling stage is the most vulnerable time in many birds' lives, and predation is their primary cause of mortality.[1] Arms races between the bird's traits to reduce predation and the predator's selection to increase it have resulted in seemingly bizarre extremes of natural history. Conspicuous examples include swifts, nesting deep in caves where they make nests entirely from strands of condensed saliva produced by the male. He winds his salival secretions into a half-cup plastered onto an overhanging rock outcrop. Oilbirds similarly have found safety nesting so deep in caves that they have evolved echolocation to negotiate passage to their nests, which are stinking heaps of regurgitated plant materials and their own droppings.[2] Cactus wrens, thrashers, roadrunners, and several other species nest inside "jumping cholla" cacti (chain fruit and teddy bear cacti), where they are protected by the cacti's spines. Puffins and petrels nest on remote offshore islands in deep burrows, and marbled murrelets, also seabirds, nest in the canopies of rain forest trees. Inaccessibility of the nest is, however, seldom if ever absolute because one adaptation begets another in the co-evolutionary arms race with the predators. As in any arms race, the best way to sidestep the conflict is through ever-greater innovation and uniqueness, aside from physical isolation. One such innovation concerns the nest structure itself. Many weaver birds weave long entrance tunnels onto their nests that hang down from the nest. These likely evolved because they made it more difficult for predators and brood parasites to enter them. Meadowlarks in North America sometimes build tunnels that also hide the nest entrance on the ground in front of their domed nests in meadows. In some species of weavers, nest construction involves suspending their nests by long ropelike extensions from a slender twig over water. The cape penduline tit builds a feltlike bag nest with a false nest entrance leading into a blind sac. The real entrance is above the false one, and like an elastic top of a sock, it closes after the bird slips in or out. The yellow thornbill similarly has a nearly camouflaged side-entrance hole into its nest, and it has a dummy nest cup on top of its functional nest. Size may also prove to be a barrier to predators. Woodpeckers drill holes into often live solid wood, leaving the smallest possible entrance. These constructions are then used for many successive years by a great variety of other birds as their own nest sites, after the woodpeckers' usually one-time use.

*Nest safety in invisibility.* A main defense of most small birds against their many nest predators is to hide their nests. And one of the means of achieving inconspicuousness is to be different from what predators are looking for, to defy the hunters' expectations of what a nest is and where it might be. Predators rely on instinct modified by knowledge gained from past experience. They scan for a particular target that they have encountered before. What is different from their repertoire of search images tends to be overlooked. For example, a predator familiar with ground nests under juniper bushes would concentrate a search there and thus be deflected from searching on branches in the crotches of maple trees, and vice versa. Since hiding depends on novelty, it is highly advantageous for birds to invent ever-new and unexpected nest locations. Large differences have evolved in different species. In the forest next to my house, for example, a wood peewee's lichen-decorated nest mimics a knot on top of a thick limb. A black-throated green warbler's nest is high in a tree fork, and the nest contours follow that of the tree to fill the notch. A winter wren's nest of green moss and fir twigs is tucked deep into the roots of an over-turned tree, while a slate-colored junco's is sunk into the ground under a cover of last year's dry grass.

Sometimes less is more. The nest of the mourning dove is a platform of a few bare twigs that hardly passes for a nest. When you see the bird on it you hardly suspect that there may be a nest under it. Fairy terns on tropical islands lay their single egg on a bare branch of a casuarina tree without any nest at all, and Arctic terns just lay their two well-camouflaged eggs in a scrape on a pebbled beach.

Since I have long been fascinated by bird nests I try to pass on the enjoyment that I got when I was a child. I recall being particularly enamored of some of the first nests I ever saw—nests of winter wrens and brown creepers and one of a long-tailed tit were especially memorable. I see them still as though I had found them yesterday. Whenever I take my Winter Ecology students in the woods and we find a tip-up, I look for one. We don't find many winter wren nests—there is perhaps only one in twenty tip-ups that has been left from the previous summer, or maybe even from the summer before that. "There is a bird's nest here," I tell them when I finally see one and they have gathered around: "Try to find it." Since none of them has ever seen a winter wren nest, they may stare directly at it without noticing it. Then suddenly there comes a revelation, and a new world opens up as their eyes connect with consciousness. Any wren nest is long used up in the winter, but it is usually still in good shape. We pick it apart, noting its outer shell of moss and twigs, and the inner lining of perhaps grouse and owl feathers—the lining that once cradled the eggs and naked young.

I may then lead the students to another spot and ask them to look again. Their eyes tend to follow every branch and twig. Still they find nothing. But I have located a tiny twig or two peeking out from under a loose piece of bark on the trunk of a dead balsam fir tree, and I know that under that loose piece of bark is the nest of a brown creeper. This bird, which is slightly smaller than a chickadee, wedges its nest into an inch-wide space under the sloping roof of bark that is starting to separate from the tree. Such nests are sheltered from rain, and the tiny nest cup is lined with shredded cedar bark, plant down, and spiderwebs, built onto a platform of loose twigs that the bird has jammed in under the bark to hold up the nest cup. In my mind I imagine the cozy safety from rain and predators that the bird must find there. Looking like a little mouse running up a tree, the brown creeper can disappear by scooting almost unnoticed under the loose bark to its nest.

The students are now on their way—they've seen two nests. There are at least thirty others in these woods that they are standing in, and it is unlikely that any of these students will get to see more than several of them in their lifetime. Not, that is, unless they go leagues beyond the usual birdwatcher stage in their development as field naturalists. Unless they are told that the purple finch is likely to build its nest in the tip top of a balsam fir tree in the densest twigs there, or that the slate-colored junco hides its nest under dead matted grass on the ground, or that the black-throated green warbler wedges its nest into the fork of a tree while the parula warbler weaves its nest into a tangle of dangling old man's beard (*Usnia* lichens), and the chestnut-sided warbler's flimsy cup of thin grass fibers is to be found in dense spirea bushes close to the ground in a clearing, and the olive-sided flycatcher's at the tip of a high branch in a spruce, they are not apt to find any of them. But finding one by chance would likely hinder rather than help their efforts to find one of another species. The nest-site variety is functional in hiding the nests from predators.

Uniqueness pays off. If the second hypothetical bird in our imaginary forest were to choose a unique place, then one would not have a clue where to start looking.

For the most part the unique nest sites are species specific. Nevertheless there is also plasticity of nest-site selection in some species. For example, eastern phoebes have traditionally nested on cliffs but now nest almost exclusively on human dwellings, which serve as substitute cliffs. They are certainly not hidden anymore, even though they are still "camouflaged" with green moss such as is found on natural cliffs, but they are now under our protection and tolerated if not welcome. Common grackles most commonly nest high in the inaccessible crowns of dense conifers—at least in urban areas I often see their nests there. However, each spring a group of them arrives in the bog near our house and at least a half-dozen pairs nest in one patch of cattails within it. Their nests are often within five or six meters of one another and are built into the wilted old cattail fronds from the year before (long before any new green shoots have come up), so they are within inches of the water. One year when all the nests were destroyed (probably by an otter or a mink), I found one grackle re-nesting in a bird box I had put up for wood ducks. Another used a birch tree in the nearby woods, and still another nested on a window ledge behind a screen of Virginia creepers. They have also been reported to take advantage of the interstices of an osprey nest as a nest site. Such plasticity may be unique. However, possibly the most extreme single example of individual flexibility that I am aware of concerns a woodpecker, the northern flicker, who on an island off New York where suitable trees were not available laid its clutch of six eggs directly on the ground in a depression in the sand.[3] Usually flickers lay their eggs in pre-existing tree holes or they excavate a tree cavity like other woodpeckers. Ospreys and bald eagles may also nest on the ground. A colony of ospreys existed on Gardiner Island near Long Island until the 1930s and still nest on sand islands of Scammon's Lagoon in Baja, California. Bald eagles nest on the ground in the Aleutian Islands, and peregrine falcons on sand, on islands off the Scandinavian coast. The flicker, however, did not just nest on the highest points available, as did these raptors, but it switched from its traditional hole nesting to having a totally open "nest." (Flickers, like other woodpeckers, may build a nest hole, but they never build a nest.)

Hole nesters may appear to have great flexibility in nest site selection, in part because cavities may be found in all sorts of configurations aside from tree holes. Perhaps they are thus predisposed to flexibility. Among the titmice—a number of genera that include our common black-capped chickadee, the perennial bird-feeder hog, along with six European species—some accept any kind of cavity they can find. The (Eurasian) coal tit and blue tits are not fussy, and aside from using old woodpecker holes they may sometimes also nest in the interstices of raptor nests, in wren, swallow, and squirrel nests, holes in the ground, and between roots in tree tip-ups. The blue tit and the great tit occasionally even build free nests with a roof. The others, such as the marsh tit, willow tit, and the crested tit in Europe and the black-capped and boreal chickadees of North America are more fussy in the cavity they will accept, and if they don't find any that are suitable, they will enlarge an existing hole or even hack out their own from a rotten stump in the manner of woodpeckers. A deep nest cup of moss and hair, and other feltlike material is then always built within the site they have chosen or made. Nest construction takes five days to two weeks. Other titmice instead construct their own elaborate open nests that resemble bags, and they show little flexibility, never using a pre-existing cavity. It is as if their expertise has been perfected and then solidified. Of these, the long-tailed tit builds its nest in a sheltered place such as the junction between the trunk of a tree and a large limb, where it tends to be camouflaged by appearing to be part of the tree. The bearded tit places a similar nest in a very dense reed thicket, and the penduline tit's nest is free-hanging, and it takes them two to three weeks to build. It has a short entrance tunnel and the nest is built over water attached to a flexible twig.

By having suggested the unique, I do not wish to forget the common and the obvious. Most birds are not as flexible as my previous examples suggest, and for every bird that may be benefited by or can adjust to human disturbance there may be a hundreds or thousands that can't. In these cases their populations are depleted, or they become rare or go extinct. Especially at risk are those that rely on old trees for nesting holes but who, unlike the chimney swift, cannot switch to using chimneys or some other human contrivance. The forest goes, and then they go and are soon forgotten and hardly missed. Who remembers the thick-billed parrot, a beautiful green bird with a crimson face and crimson feathers on its shoulders and legs? In Roger Tory Peterson's 1941 *Field Guide to Western Birds* it was described as a visitor to the mountains of Arizona and New Mexico. It visits there no more. This parrot feeds on pine seed, and although pine trees were not the limiting ingredient to its populations, cavity-ridden old pine trees for nest sites were. Several pairs would live in a semi-community in one cavity-riddled old tree.[4] Those cavities in old-growth pine forests used to be provided by the imperial woodpecker (an extinct relative of the ivory-billed woodpecker). Even birds that were still common within recent memory are at risk because of lack of nesting sites. In portions of the United States and Canada the American kestrel, who used to be ubiquitous, has in recent years experienced what was described by the 2008 *Pacific Raptor Report* as "precipitous" depopulation. I still occasionally see kestrels every spring in Vermont, but untidy old barns with haymows and large hollow trees at the edge of old pastures are infrequently found now, and I seldom see one nest at those sites that used to be available almost everywhere. In Germany, despite modernization, nesting places have been created by designing nest sites in new buildings to accommodate kestrels, jackdaws, swifts, and other birds, and these have had sometimes dramatic effects on the species' populations.

*Nesting neighborhoods.* One important criterion of especially long-lived birds' choice of nesting spot is habitat—surroundings that support their lifestyle, yielding sufficient resources to feed the young, providing relative safety from enemies, and offering a place to meet potential mates. Having been born at a place or having nested there successfully is the ultimate proof that it is on average safer to nest there again than to nest in an unknown place. Others may return to the general area where they were born and settle in the vicinity, if not the spot vacated by a deceased parent. Canada geese have a strong tendency to return to their birthplace. The two female geese that I had reared each came back home to nest, but they faced stiff competition and were driven away. The home area may be preferable, but it can be crowded and dispersal is necessary. There may even be a large payoff in colonizing a new area, and wandering and looking for new locations can be a necessary alternative to staying at home. If one could identify individuals, one would see many strangers passing through who are all looking for a place to nest.

Rare birds (and therefore a unique voice and presence in a neighborhood) allow one to learn something about movements and settling. In the spring of 2008 a golden-winged warbler showed up singing near our house. I had never before heard or seen one. This one stayed for four days, and every day in the morning and in the evening he sang, switching between two locations separated by about three hundred meters. And then on and after the fifth day there was no song at all. I suspect he had left to try to attract a mate elsewhere.

Like last year, a wood peewee started calling in the woods in the back of the house starting on June 2. He sang almost continuously every day until June 14, after which he was almost but not totally silent. In this case I presume he had found a mate or had started nesting. And for three years a mourning warbler sings every day for several weeks from the same branches on a birch tree above an area of brush and brambles in front of the house. In the last four years a male bittern has come to the same marsh in the third week in April. For about a month he calls loudly every morning and evening, and then abruptly there are no more calls at all. I have never seen any young and have been unable to find a nest, even though I have searched long and often, so I think it unlikely that a mate came. However, he stays and invests the whole spring in trying to get one, perhaps because he has found a choice spot.

Many of these birds vocally advertising from almost the same spot may be the same individuals. Most birds greatly reduce their singing after they have a mate, and so I presume their song durations indicate either how long it takes to get a mate, or how long they are willing to advertise for one on the same potential territory. Habitat quality in part determines how long they would advertise. However, many birds settle preferentially near others who may indirectly be habitat indicators.

It struck me as odd that the least flycatchers, which I had seen only rarely in the last several years, suddenly seemed common in 2007. There were several pairs of them in the surrounding woods just below the small rise where our house sits. The following year, to my great surprise, there was not a sound from them there. At first this alarmed me, but then I discovered that they were common at another spot about two kilometers down our dirt road. Least flycatchers are never conspicuous, but in this area I heard them "everywhere." I stopped there almost every day to hear them sing at the same places so that I could plot approximate territory sizes. I estimated there were at least six pairs along a stretch of two hundred meters. I saw frequent chases. After about a half-hour of searching I found two nests. I wondered, are some birds organized into neighborhoods? H. H. Harrison in his detailed observations of wood warblers (Parulidae) already concluded that the normally territorial Wilson's warblers neverthe-less nested in "colonies," although he presumed that the reason was simply habitat.[5]

Recent studies support the idea that although those birds who had nested successfully the previous year return to their home terri-tories, others respond to cues from those who have already settled in what is presumably previously established suitable habitat. Playbacks of American redstarts' songs in random plots both early in and dur-ing the breeding season recruited more males new to the area, al-though yearling males, who arrive about a week later, failed to re-spond.[6] The reason for attraction to the songs may have been related to territorial considerations or interactions with neighbors, but a sub-sequent study suggests that social cues are also used for settlement.[7] In this study of the black-throated blue warbler the researchers broad-cast song playbacks in habitats where this species normally would not settle. Young males without breeding experience were recruited there, even though the playbacks were broadcast the previous year after the breeding season, so apparently the experimental cues could override habitat. The authors suggest that males without territories prospect for potential territories even before they leave on migration. Most likely they then remember the locations and migrate to them the following spring. Females, who return from their migration after the males already sing in established territories, may simply travel through an area or fly over it identifying good places to nest by listen-ing for the males' advertisement songs.

*The safety of the neighborhood.* One of the conspicuous behaviors of many birds is their mobbing of predators, nest molesters, and nest parasites. The attraction to others who are excited or agitated may appear to be courting danger, but instead it functions as a relatively safe way to learn what is dangerous from another who has knowl-edge or experience. The knowledge is then passed on and becomes culture. As such it pays to be near others, and with respect to nesting it helps to have neighbors. Density can vary enormously. Some gulls nest in dense colonies where a community defense is possible. Other birds live semi-colonially, such as in blackbirds. Still others form very loose aggregations, such as in least flycatchers.[8] The "neighborhood" may be a common phenomenon that is like a greatly extended colony. Locally in Vermont I have seen marsh wrens nearly everywhere in two cattail marshes, but they are absent at other marshes that to me look identical. Similarly, I had found six or eight gallinule nests in one marsh but have not seen one anywhere else at any other marshes in Vermont or Maine. About fifty years ago bobolinks seemed to be plentiful in every hayfield. Now most hayfields are barren of them, although any field where a bobolink still sings has several of them while a neighboring field has none. Habitat, as such, is the main predictor of where to find the nests of each species, but not all suitable habitat may be filled as more species become rarer.

Red-winged blackbirds are still numerous enough that they occupy almost all suitable habitat in large numbers. Although they are not conspicuously "colonial," they breed in neighborhoods that could be classified as loose colonies. In this case the neighborhood is probably the immediate result of clumped resources—the nesting sites and food in marshes. The icterids have a wide spectrum of gregariousness. Almost all are highly gregarious most of the year when they are not nesting, and during the nesting season they may nest one next to another (as in oropendolas), in loose aggregates (North American blackbirds and grackles), and seemingly alone (as in meadowlarks, who are nevertheless often polygynous).

There is a similar spectrum of nesting in the swallows. Most tend to be gregarious, and part of that tendency to nest in the neighborhood of others is probably related to localization of nest sites—caves or barns for barn swallows, sand banks for bank swallows, and cliffs or equivalents for cliff swallows. Barn swallows sometimes nest singly and often several together in loose colonies. A colony of over fifty pairs of cliff swallows nested for many years under the roof overhang on the outside of a barn near Wilton, Maine, but there were none at other barns then, and for decades now there have been none at the former site. Similar nest sites, however, are practically everywhere.

How do these neighborhoods of breeding birds originate? I speculate that it could be a similar phenomenon to that of some amphibians in which the females are attracted to the place where several or many males are advertising because several males simply provide more signal. When one tree frog calls, others nearby join in, and a female therefore can be attracted from a longer distance. She then has the added advantage of having more choice of mates because she can compare. All gain by added predator protection.

Once the area has been decided on, there is the decision of where to build the nest. In most birds it is the female who picks the precise spot where the nest will be built. She lays the eggs and prepares the place for them, so she makes the ultimate choice. But if nest sites are rare, such as nest holes, then males can help in finding and securing sites. In those cases the male who finds a good nesting spot could potentially also use it to entice a female to come with him so that she can choose it, and him. Alternately, the pair can form first and then travel and house-hunt together.

Many birds, such as the geese, ducks, and starlings mentioned previously, seem to be already paired when they go house hunting. Perhaps the male starling finds many potential nest holes and brings his mate to them, because he inspects a potential nest hole and then perches directly next to it, expressing his enthusiasm for it by singing his energetic nonstop jazz improvisation, mimicking sounds and songs he has heard at some time in his life. All the while he flaps his wings in rhythm with his song. The female then joins him and again and again they inspect the nesting place he is advertising. Then they leave for hours or days, possibly looking for more attractive sites else-where. I presume it is she who ultimately chooses the nest site be-cause she alone builds the nest. She alone incubates the four to six light blue eggs at night, though he takes turns in the daytime. Others, such as storks, eagles, ravens, and some hawks, swifts, and swallows also return to the same nest sites or nests.

A male sandpiper that has just returned to the Alaskan tun-dra performs a spectacular flight display accompanied by conspicu-ous and distinctive vocal repertoire to attract a female. As a bird who needs to fly nonstop for thousands of miles in order to reproduce, flight enthusiasm may reflect flight capacity, ritualized into a sexual display that accurately reflects reproductive potential. Having gotten the attention of a potential mate that follows him around, he makes several prospective nest scrapes to signify his intentions and his choice of her. She signifies her choice of him by inspecting the nests and choosing the one that suits her. In it she will install a few bits of grass or sedge, and then every morning for four days deposit an egg.

A male weaver bird in a breeding colony makes a single nest hull that dangles from the tip of a slender branch. He then hangs from it with his feet, flutters wildly, and vocalizes whenever a female is nearby, much as a male starling advertises a nest hole. In the weaver colony, however, there are other males doing the same nearby, also vying for a potential mate to come and inspect their nests. The dis-cerning females choose among the males and their products. After joining a male or his nest, the female finishes it by adding the lining and then lays their clutch of two or three eggs.

In contrast to the weavers, a male marsh wren commonly builds six or sometimes even seven or more nests in his territory, and then he sings vigorously most of the day until a female is attracted. She inspects the nests, and presumably the more there are the longer she stays around, perhaps finding one to her liking. His building activ-ities are, like the woodcock's aerial display, an honest advertisement of his energies—energies that are based on qualities such as foraging skills, health, digestive capacities, and vigor, which generally translate to increased help to and viability of his offspring and the offspring of the mate who chooses him.

Aside from one sex deciding where their mutual nest will be is the issue of species preferences and constraints that determine the features of the precise nest construction and locations that are suitable.[9] Chipping sparrows build their nest cup lined with long thin hairs in the dense foliage of a conifer. The red-eyed vireo suspends its cup nest in a horizontal fork of a deciduous tree. Least flycatchers build theirs into the three-tongued fork of a slender tree; a golden eagle builds onto a high crag or a tall, sturdy tree. And a northern oriole hangs its bag nest from the thin twigs at the end of a long, freely swinging branch. But aside from such specific geological and botanical features in part dictated by habitat, some birds choose to nest in the nests of other birds.

*Nests in other nests.* Given suitable habitat, the nests and nest sites of one species are often suitable for another. Of probably greatest use for a number of diverse species of birds are the nest holes excavated by woodpeckers. Woodpeckers themselves don't build nests, unless the wood chips remaining on the floor of their nest holes count as a nest. Their excavated cavities are generally used by them only once, but they provide sites wherein other birds may nest for as many years as they may exist afterward. Some species of weaver bird, thrush, flycatcher, titmouse, nuthatch, swallow, wren, warbler, crow, swift, and starling then build their own nests inside these cavities. Others, such as some species of owl, falcon, barbet, honey-guide, kingfisher, and duck, accept them without adding nest material. In Europe, for example, nest holes excavated by the green woodpecker are taken over by a progression of different bird species over the decades that the nest hole exists. First comes the nuthatch, which applies mortar to make a smaller entrance hole, then in subsequent nesting seasons come great tits, blue tits, and redstarts. Much later, as decay enlarges the hole, comes the wood dove and the tawny owl. When the larger black woodpecker excavates, the progression would start out with the latter two species, and with smaller woodpeckers the succession of occupants would start with titmice. In Africa, in addition to wood-peckers there are weaver birds who provide woven nests rather than cavities that are later used as nest sites for others.

The mud constructions of swallows could arguably be con-sidered nest sites analogous to woodpecker holes and bird houses rather than nests as such. Like woodpecker holes, they provide a suit-ably protected place where a nest of grass lined with feathers can be built. As with tree-hole nests, a variety of birds may use the swallow mud "nest" structures within which they build their own nests. In Texas, cave swallows have expanded their range apparently because they have discovered and modified the nests of barn swallows, and thus they now colonize new areas where they were absent before.[10] In Alaska violet-green swallows will nest in cliff swallow nests, and a solitary sandpiper was seen nesting in a robin's nest (J. Wright, pers. comm.). Some species may take the swallow nest even before the swallows get the chance to move in themselves. House sparrows, who commonly take over barn swallow nests, are perhaps best known for this behavior. Normally they build their typical domed nests with a side entrance resembling that of many other weaver birds out in the open or in dense cover, but cavities provide the best cover. Similarly, other "sparrows" like our common house sparrow (also a Ploceidae or weaver) in Africa (including *Passer swainsonii, P. gandamensis,* and *Sorella eminibey*) build their own nests, but sometimes appropriate the nests of other weavers. In mixed-species weaver colonies, where the birds fight for and take over the woven nests of weavers, it sometimes occurs that the weaver nest builder is already laying eggs. Rather than driving him off, the "sparrow" lays its eggs in the nest and the weaver then rears the sparrow's young.[11]

Very large nests, such as those of raptors and the communal nests of the sociable weaver of southern Africa, are used as nesting platforms for many birds. They become nesting sites and a magnet for other bird life. The South African pygmy falcon is almost completely dependent on the sociable weaver for its nesting sites.[12] In addition some barbets, chats, finches, a tit, and lovebirds also use the sociable weaver's nest chambers, while the roofs of the huge nests become nesting platforms for owls, hawks, eagles, and even a goose. Raven and crow nests often become nest platforms for hawks, owls, and falcons.

*Association with raptors.* Raptors are potentially good bodyguards because they prey on or chase animals that could be nest raiders of their own and others' nests. Snowy owls, for example, attack Arctic foxes that threaten their nests, and snow geese gain protection by nesting near them: of 236 snow geese pairs, 90 percent nested successfully when they were no more than a kilometer and a half away from a snowy owl nest. When no snowy owls were near, they had only 23 percent nesting success.[13] Many birds nest even closer to raptors. Ravens commonly nest on the same cliffs with peregrine falcons, and many smaller birds even nest directly in the interstices of the raptors' nests.

The huge nests of large raptors, although they cannot be hidden, need not be. On the Mongolian steppes, golden eagles nest on the ground in flat open country. A nest can be seen for miles. It is made up of the accumulated debris the pair has collected over the years, and it looks like a column from the distance (R. Opalca, pers. comm.). There are wolves there, but the female eagles, 15 percent larger than the males, are formidable enough to be used by people to

hunt wolves. A Mongolian horseman will release a female golden eagle from his arm to attack a running wolf. She strikes the wolf by
clasping around its mouth and holding it until the horseman arrives
and kills it. It is doubtful that a predator would be willing or able to
raid the eagles' nest. In turn, other birds can use the eagle nest as a
nesting place.

In his book *Raptors*, Scott Weidensaul notes, "A raptor's nest
would hardly seem to be the safest place for a small bird to raise its
own young, yet birds have frequently been observed nesting inside
the bulky mass of a large hawk or eagle nest."[14] He notes that in the
American West kestrels and kingbirds sometimes nest in the interstices of active golden eagle nests, while orioles, kingbirds, and house
sparrows use the nests of Swainson's hawks, and magpies those of
ferruginous hawks. On the East Coast of North America, on the
other hand, house sparrows, house wrens, common grackles, and
sometimes even black-crowned night herons build their own nests in
one currently used by an osprey. Even accipiter hawks, as Weidensaul
points out, despite their specialized bird-hunting lifestyles, "will not
hunt in the immediate vicinity of their nests, so songbirds living close
by may be safe from that threat, as well as from other predators like
crows that the hawks would drive away."

Raptors kill for a living and have therefore evolved strong inhibitions against harming anything at or near their nests because it
could possibly be their own young. Presumably this is similar to inhibitions most birds have against discriminating against eggs in their
nests, even those that may not be their own. Weidensaul cites bald
eagles who raised red-tailed hawk young in their own nest.[15] The
young red-tails had apparently been brought into the nest as prey, but
since they were still alive when they got there they were then adopted.
Another perhaps similar example involving the same mechanism is a
glaucous-winged gull chick that was reared by bald eagles in their
nest. Near Kiel, in northern Germany, I had the good fortune to ac-

company Thomas Grünkorn, a conservation biologist, to see a sea eagle nest that had long been under investigation. It not only regularly fledged sea eagles, but the huge stick nest was also a nest site of starlings, house sparrows, tree sparrows, and wagtails. I have also seen house sparrows nesting in a Swainson's hawk nest in Caldwell, Idaho.

Canada geese sometimes use hawk nests, probably because elevated platforms are safer sites than ground nests. They may even drive an osprey pair away from an active nest to take it for their own, though in Vermont the geese most commonly nest on active and inactive beaver and muskrat lodges that are surrounded by water. In Europe one sees conspicuous and often large colonies of rooks in the middle of towns where there are trees. Numerous falcons and owls use as their nest platforms the abandoned nests of other raptors, ravens, crows, and magpies. Magpies have moved into towns, and jackdaws, who have traditionally nested in ruins, live in town as well, although with prosperity leaving fewer ruins and crumbling buildings, some jackdaws have started nesting in rook nests.

In some situations even raven nests can become nest sites for other birds, even though one of the ravens' preferred foods is any and all birds they can catch. On a trip by camel across the Libyan desert, biologist Herbert Biebach encountered a lone acacia tree that had a common raven's nest on it (H. Biebach, pers. comm.). He climbed the tree and found three feathered young inside the nest mold. However, the twig understory of the nest held several active nests of the desert sparrow. Ravens I have observed sometimes catch adult birds and seemingly never pass up eggs and nestlings if they can find them. The presence of potential prey directly at the nest may seem miraculous. However, I've seen ravens incubate fresh hen's eggs placed into their nests, although outside the nest such eggs are prized food.

*Association with humans.* We are enemies of many birds because of habitat destruction. However, locally some birds also use us as allies.

Swallows of many species are now almost dependent on humans. In the northeastern United States, the open fields and pastures provide the foraging habitat for at least four species of swallows. They perch on electric lines and build their nests (depending on the species) in and on our buildings, in bird houses we provide, and in gravel pits. They follow the mowers and harvesters over fields to catch flushed insects, just as seagulls do. Phoebes seem to have taken advantage of human protectors, who are often not averse to getting rid of marauding chipmunks and squirrels. There are almost no more Eastern phoebes "in the wild" nesting on cliffs. The most common of local birds, American robins, once nested only in the forest, but some of them are increasingly cozying up to humans. Robin nests, when they are located in deep woods, are one of the easiest bird nests to spot. I find at least a half-dozen every summer along the wooded roadsides where I jog regularly, but seldom do more than one or two of them survive long enough to fledge young. Most get raided during the egg-laying stage. Again and again I see a nest built, the pretty blue eggs laid, and a day or two later the nest is empty. However, those robins who now build their nests on and in structures such as barns and houses are spectacularly successful in bringing off young.

Many of our neighbors hang decorative wreaths made of vines or twigs on their doors, and they commonly become a robin's nest site. Here the nests almost always fledge young. Our neighbor's sometimes fledged three broods in one summer, and so I also put up a wreath on our garage door. It immediately attracted a robin pair who successfully raised two broods in succession in the same year. I noticed another robin nest on a beam along a walkway in an active shopping mall, and the late biologist Vincent Dethier in his book *The Ecology of a Summer House* described a robin pair that successfully reared their young on an active ferry on the Maine coast.[16]

Wood warblers (Parulidae) are very specific in their choice of nest sites, and only two, the Lucy's warbler (who lives in deserts) and the prothonotary warbler (who is at home in swamps), nest in cavities. The prothonotary warbler typically nests in a rotting stump cavity over water. But apparently what a prothonotary warbler considers relevant for a nesting site is a cavity to tuck its nest into that is near or over an open surface, which pre-adapts it to all sorts of human-created sites, such as a lawn. As Harrison describes, it readily nests in birdhouses and in hollow gourds that are hung up. Its nests have also been found in "an open window, a mailbox, a coffee can, a cheese box, an enamel coffee pot in a gazebo, a hornet's nest, a large glass jar on its side, the pocket of a pair of pants hanging on a clothesline—and a pulley wheel in a ferry boat that was in operation from one side of a river to the other."[17]

In North America, barn swallows, chimney swifts, Eastern phoebes, house sparrows, and rock doves are so habituated or adapted to humans that they now rarely nest in "the wild." Others are not such obligate human associates, but they seem to be associating more closely with us. In Europe they include about two dozen bird species that are moving ever more out of the wild and into towns and cities. They range from small songbirds to most of the corvid species to some falcons, ducks, and geese. All are taking advantage of the food we provide directly or indirectly and the nesting sites that are nearby. Nonmigratory house sparrows have colonized northern Canada where temperatures may reach −50 °C, chiefly by using grain elevators to find both food and shelter.

We have the power to eradicate animals that are attracted to our food and dwellings, and often we do, but on the whole we become tolerant if not fond of them through familiarity. Our dwellings have features in common with cliff sites, and cliff-nesting birds such as peregrine falcons, ravens, and red-tailed hawks use high-rise apartments as substitute cliffs. "Pale Male," for example, is a red-tailed hawk who famously arrived in New York City in 1991, where there are now about fifty pairs.

On a visit to the Public Museum in Milwaukee, Wisconsin, I was hosted by Greg Septon, who had worked for about thirty years on peregrine falcon recovery. I stayed at a downtown hotel, and from my fourth floor window had a view of the forty-first floor ledge nest site of "Sibella" and "Bill." Sibella had been bred in captivity and was released in 1988 at Isle Royale National Park. She chose to nest not there in what to us might seem to be Edenic surroundings but a high-rise building in Milwaukee in 1989. She had come back to it every year ever since, and she and her mate raise a brood of about four young. (Bill is her second mate.) Other peregrines came as well; Greg pointed out several other peregrine nest sites on other buildings.

Initially attempts had been made to re-introduce the peregrines at their traditional nest sites along the Mississippi and Wisconsin rivers. But there the nesting always failed because great horned owls lived there, and they destroyed the young falcons. Owls were shot to try to protect the falcons, but new owls invariably replaced them. However, after deliberate release of falcons in the city, these birds experienced a phenomenal (nearly 100 percent) nesting success, and their population "took off." In Milwaukee alone there are some twenty nesting pairs of peregrines that raise about sixty young per year. They nest not only on "natural" building ledges but also use artificial ones provided specifically for them on such buildings as the Kenosha Medical Center, the Miller Brewing Co., the Pat Washington Power plant, and on platforms attached to smokestacks along the Lake Michigan shoreline. As the peregrine population has increased, competition for the favored sites has become fierce, and some birds are starting to nest again on their less-preferred natural cliffs, despite the danger of predation by great horned owls.

Like other birds, falcons return year after year to their successful nest sites. Furthermore, the young appear to become imprinted on the type of nest site where they grew up. Fledging provides the young the ultimate proof of their parents' correct choice of nest site—one where they themselves are likely to become successful parents. Successful nest sites thus become traditional and then become well known if not loved by their patrons. A mallard duck nested annually in a planter thirty meters up in the Kiel, Germany courthouse, and each year the newspaper announced when her young would jump down and be led by her to safety, crossing a busy street to reach water.

Imprinting on a successful nest site resulting in the establishment of a tradition may also be relevant in other birds. Ravens in the western United States, the Alps, and Vermont nest primarily on cliffs. In Maine, and in Poland in the Bialowieza reserve of forests with many ancient alders, oaks, beeches, spruce, and basswood, ravens nest primarily on pines. In northern Germany, they nest primarily in the forks of tall beech trees. In some cities, including Kiel, Toronto, and Milwaukee, both herring and ring-billed gull colonies have become established on the flat rooftops of city buildings, and as city rooftops have been getting "greener" by the planting of grasses to "provide habitat in the clouds," they may become preferred nesting sites.[18] These sites are free of predation by foxes, weasels, otters, and raccoons, not to mention the disrupting effects of bathers, doggies, stray cats, dune buggies, and motorized traffic. All over Africa, where snakes are primary nest predators, weaver birds place nests at the tips of branches hanging over water, village squares, and other public places. Snakes stay clear of humans, and those who don't generally have a short life span, especially in the tropics where many are venomous.

Our more modestly sized dwellings are, to many birds, also ersatz cliff faces for nesting. Swifts and swallows are perhaps the most frequent users of our dwellings, and in Europe the house martin and in America the barn swallow are almost completely dependent on our structures. I have never heard of either one nesting anywhere but on human dwellings. On the other hand, the cliff swallow still nests commonly on both buildings and cliffs.

No other type of human dwelling has historically been such a haven for birds as our traditional barns, such as the barns I knew as a youth in rural Maine. They were huge, cavernous structures with horizontal hand-hewn beams holding up the roofs, with cow and horse stalls, chicken pens and pigsties, and with holes under the roofs and under the eaves. The wide, open doors were large enough for the hay wagons to be pulled in by the tractor or a team of horses, and in and out of those doors flew the flocks of barn swallows nesting on the ceiling beams. Starlings and house sparrows and occasionally a kestrel nested in the crevices, and the possibility of a barn owl existed as well. A pair of Eastern phoebes and robins nested generally under the barn where the manure was hauled out. Chimney swifts cruised over the adjacent house and in the evening dropped down into the huge brick chimney. The farmyard provided food—there were grain scraps from the animal feed and seeds from the hay. It was a breeding place of flies and mice—and there was safety, and to me also a fount of fascinating life. One of our neighbors in Maine had a colony of over fifty cliff swallow nests plastered onto the outside of his dairy barn. In Europe such an untidy and partially dilapidated structure would have been graced with jackdaws—small gray-necked crows—as well.

Chimney swifts used to nest in large old trees that with age become hollow. In New England such trees were primarily white pines. Further south they would have been sycamore trees—one of which Johnny Appleseed camped in for awhile. They provided roosting and nesting places. Now such trees are no more, but Donald Culross Peattie in his tree book, *A Natural History of Trees in Eastern and Central North America* gives us a vivid picture of what is lost:

> Long before there were any chimneys to send up a twirl of smoke in lonely clearings, these hollow sycamores were home to the chimney swift—"swallow," as the pioneers often called it. On an evening of July, not far from Louisville,

Kentucky, there came to such a tree John James Audubon. "The sun was going down behind the Siver Hills," he remembers, "the evening was beautiful; thousands of Swallows were flying closely above me, and three or four at a time were pitching into the hole, like bees hurrying to their hive. I remained, my head leaning on the tree, listening to the roaring noise made within by the birds as they settled and arranged themselves, until it was quite dark when I left. . . . Next morning I rose early enough to reach the place long before the least appearance of daylight, and placed my head against the tree. All was silent within. I remained in that posture probably twenty minutes, when suddenly I thought the great tree was giving way, and coming down on me. Instinctively I sprang from it, but when I looked up to it again, what was my astonishment to see it standing as firm as ever. The Swallows were now pouring out in a black continuous stream. I ran back to my post, and I listened in amazement to the noise within, which I could compare to nothing else than the sound of a large wheel revolving under a powerful stream. It was yet dusky, so that I could hardly see the hour on my watch, but I estimated the time they took getting out at more than thirty minutes. After that departure, no noise was heard within, and they disappeared in every direction with the quickness of thought."

Since Audubon made his observation in July, the breeding season of the swifts, it is likely that he had observed a colony. After the pioneers cleared the forest of large trees the swifts took to their chimneys. The wide chimneys from wood-burning fireplaces are almost gone now. The new houses have narrow metal tubes sticking out of the top to vent gas from oil burners, and I seldom see a chimney swift or a cliff swallow any more. Barn swallows are becoming rare. I used to watch the barn swallows nest in our university barns until recently, but when I last checked I saw nets strung up under the roofs to "keep out the birds." They indeed totally excluded the barn swallows. But the rock doves ("pigeons") still came to feed (they were now nesting in the city of nearby Burlington). However, house sparrows and starlings were still nesting there. They used old rat holes chewed into the walls and these provided entrance to ideal nesting sites for these hole nesters.

*The ant guard.* If nesting near potential enemies such as raptors and humans that act as guardians to create the "dear enemy" effect— where a potential enemy is recognized and eventually is tolerated— then we should also see examples of birds nesting near other noxious animals.[19] Colonies of stinging ants, bees, and wasps that attack their predators would expectedly repel the birds' predators as well. Indeed, already over a century ago, P. H. Gosse noted that many tropical birds nest near colonies of aggressively stinging Hymenoptera rather than avoiding them. He speculated that they did so in order to be protected by them.[20] A large literature on the topic is now available, and at least a hundred species of tropical birds have been reported to nest within one and a half to three meters of stinging ants, bees, and wasps. The insects often attack the birds initially when they start to build their nests, but they gradually become habituated to their constant presence as neighbors, and the birds then also learn to avoid provoking their hosts. The insects only rush out of their nests to attack any new and to them unfamiliar disturbance, "motivated" by their own evolutionary logic of protecting their own valuable brood.

Ants do not disdain dining on baby birds. It therefore seems counterintuitive that some bird would nest near an ant nest rather than distancing themselves from it. Nevertheless, many birds do cozy up to noxious ants, and their effective use of an ant guard is well documented. The tropical ecologist Dan Janzen, who has studied the association between ants and acacia trees in Costa Rica, reports that orioles (Icterus spp.) and kiskadees take advantage of these ants as a nest guard.[21] The ants at first attack the birds but after a while become accustomed to them and later ignore them. The ants attack climbing opossums and snakes that raid bird nests, and the birds' nests with an ant guard have greater nesting success than those without one. Wasps, being generally even more aggressive than ants, could be potentially even better guards.

*The wasp and bee guards.* Most wasps have formidable stings and furthermore can attack more quickly and from a greater distance than ants. Some tropical Polybia wasps are extremely aggressive in defense of their nests and attack humans who come within five to ten meters of them. A study by Neil G. Smith in Panama showed that yellow-rumped caciques and Wagler's oropendolas fledged more young when they were near wasp or bee nests than when they were without these insects nearby. However, in this study the main advantage of nesting close to the social insects was apparently reduced mortality from parasitic larvae of botflies, who deposit their eggs or living larvae on bird chicks. The larvae then burrow into their flesh. A chick infested with ten maggots usually dies, and fly-caused mortality in colonies unprotected by being near wasp or bee nests is high (90 percent). Bees and wasps protect against this menace because they detect botflies that come near by odor and wingbeat noise. They then behave as though these insects, while trying to parasitize the birds' young, are any of the countless insect parasites that invade *their* nests. (In this case, in a curious twist, it pays for the host birds to be parasitized by the giant cowbird because the cowbird chicks preen their nestmates and remove botfly larvae).[22]

I was first introduced to the wasp guard when I was on a trip to Costa Rica, where I photographed a nest of the yellow-olive flycatcher. The nest was conspicuously hanging out on an open limb next to a nest of vespine wasps. I doubted that the close proximity of the two nests was accidental, and so I wrote to the ornithologist Frank J. Joyce, who made the species identifications and informed me: "Yes, this occurrence is moderately common." At least a hundred species of primarily tropical birds nest with nests of wasps and bees. He enclosed a reprint that included about thirty-five references to the topic. It covered his own study examining nesting success of rufous-naped wrens with ants who build their nests in both ant-acacia trees and near wasp nests in these trees where they are protected from predation by monkeys.[23] To find out if the wrens gained protection by nesting near wasp nests or merely by being in an acacia tree, he moved active wasp nests to randomly chosen spots in the trees with wren nests. He found that wrens whose nests were near wasp nests were twice as likely to fledge young than those who had no wasp nest near them.

African weaver birds (Ploceidae) have a variety of nest defense mechanisms that also include wasps. Some weavers suspend their nests by a long grass "rope." Others nest specifically in trees with wasp nests. Paul Beier and Agba Tungbani examined the nesting cycle of the red-cheeked cordonbleu finch, whose nests more than half of the time are less than a meter from those of the wasps.[24] They documented only four nest predations on 122 nests associated with the wasps, and eleven cases of predation on ninety nests not associated with wasps.

*Wasp nest mimic?* The forests of New England do not have nearly as many stinging and biting social insects as in the tropics. Furthermore, most that live here do not have their nests in the open on trees. There is one exception, though, the bald-faced hornet. These very aggressive wasps nest out in the open in brush and on tree branches (and sometimes under roof overhangs). They attack any intruder, and they have a hair-trigger response. Yet I have never seen a bird nest near a bald-faced hornet nest. However, the red-eyed vireo, one of the most common birds of the New England woodlands, may take advantage of these wasps in the defense of its nest in a clever ploy that relies on deception.

The vireo nest cup is attached to a horizontal fork in a slender twig, where the lip of the nest cup is a wraparound affair of birch bark and spider- or caterpillar webs. The nest is cushioned on the bottom with long thin fibers (often grass or the long, slender, dry white pine needles). It seems to be elegantly functional, except for one anomalous and intriguing feature: almost invariably the exterior is decorated with bits of paper filched from a bald-faced hornet nest. Blue-headed vireos, who nest nearby in the same woods, but on conifers (in Pennsylvania they nest on deciduous trees), only occasionally decorate their nests with wasp paper, and they apply the decoration as if it were an afterthought, only after the eggs are laid.[25]

Seen from a distance, a vireo nest in June looks like a young bald-faced hornet nest at that time because of its shape, size, and the way it hangs. However, when seen through a partially shielding lattice of leaves, the imagination paints it to be even more vividly like a hornet nest, if one has had the experience of having provoked the occupants of a real one. Since the hornet nest paper serves no structural purpose in a vireo nest, could it serve a defensive function by reducing the chances that a potential nest predator would approach a nest to discover what it really is? Fear of wasps would be required for this to be effective. I don't know what better proof is available for that than the fact that there is a syrphid fly, *Spilomyia sayi* (and four other flies and a wasp as well) that so accurately mimics a bald-faced hornet that most humans on seeing one of these flies would not suspect that it is not a hornet. Undoubtedly birds have been making the same mistake for millions of years, or else flies would not have evolved the close resemblance. We have no problem, however, recognizing what comes out of a hornet nest after having jiggled a branch that has one hanging from it. Neither would the birds.

8

NEST MATERIALS

AND

CONSTRUCTION

IT IS THE FIRST DAY OF APRIL. SNOW-

drops poke their nodding white heads through the snow and

the first crocuses bloom on snow-free patches of our yard. But

it still freezes enough at night so I can walk on the crust in the

morning before the rising sun softens the snowpack. And on

this still, dark dawn I hear a winter wren's resounding song ring-

ing from the woods. I walked out on the crust to check on our

recently returned broad-winged hawks, find their four old nests from previous years, and see the new one of this year, already up; I find a twig of fresh hemlock ten centimeters long (nest-lining material) beneath it on the snow. There is no hemlock tree within sight, so this twig must have been carried here from afar. Satisfied, I return to our house, where chickadees inspect a nearby nest box and a tufted titmouse hops on a patch of bare ground by the woodshed, picking up stray dog hairs. But although the birds are responding to the recent gorgeous days, the winter is not yet over. Two days later a storm deposited over a half-meter of fresh snow.

I had been hearing a dull but steady hammering at a broken-off poplar tree near our house where a pair of pileated woodpeckers is preparing to nest. But by April 15 their work has stopped, and they sound off in maniacal laughlike calls. Perhaps, finally, after nearly a month of work, they are laying their eggs.

Just six days later a pair of blue jays comes in the morning and perches in the birch by the house. The male takes food from our bird feeder and presents it to the female. She opens her bill and begs imploringly while fluttering with her wings as fledglings do; he is being a "parent" to her, showing that he can be one to their kids. She then hops to the ground to a patch of ground I have dug up near the garden and makes soft, fast-repeated, and almost whinnying comfort sounds as she alternately picks up twigs and debris and again discards them. Eventually she keeps a small root, which she, with him following, carries off into the woods. She is already lining their nest, which I later found in a little spruce. Within minutes they are both back, and she collects another rootlet. I then hear incessant, soft breezy-wheezy calls of a black-capped chickadee. These and other titmice make these vocalizations only at the beginning of their nesting season. She is almost frenetically foraging and never ceases her calling for a moment. Her mate, who is silent, tags along. Occasionally he finds a small moth or other morsel of food and brings it to her, and like the female blue jay, she then also vibrates her wings in baby-bird fashion while peeping softly as he puts the food into her open mouth. She is getting supplemental feeding that she undoubtedly needs to invest in her half-dozen or more eggs developing and now growing in her body. I doubt that she is already laying eggs today, though, because she stops at a cedar and starts pulling off tiny strips of bark, one after another; she is still putting lining into the nest. Occasionally she puts all the strips she has already collected in her bill down beside her onto a twig so she can rip off another one. Then she picks up the whole bunch and flies off. He follows closely behind her; he is already mate guarding since she is now fertile and he could be cuckolded.

A half-hour later I meet the chickadee pair (or another pair) again by locating them from the female's incessant calling. This time they fly inside our woodshed, and she emerges with long white deer hairs hanging out from each side of her bill so that she looks like she has a giant white mustache. Again the pair takes off, flying in a straight line into the woods where the leaf buds are now thickening and the browns of last year's dead vegetation contrast with the luminescent yellow-green of moss hummocks. Usually these chickadees nest in a cavity in a birch tree that they excavate less than a meter above the ground. I am amazed that they accomplish this with their tiny delicate bills. They can only excavate partially decayed wood, but how do they know if the wood inside a birch tree may be decayed, since they first have to penetrate the tough bark?

*May 5, 2007.* I was outside by 7:00 AM on this wind-still dawn, hearing woodpeckers drumming in all directions—the downy, hairy, sapsuckers, and even the pileated (who are now incubating). The first yellow warbler returned today and is singing in the bog. Meanwhile, a female robin makes several trips in rapid succession, carrying old wet leaves and grass and flying to and from the nest she is building on a wreath of grapevines hung on our garage door. Her mate perches and sings from the red maple tree that arches over the house. Our phoebe also started nest building yesterday; she has deposited a ring of mud for a nest foundation on a narrow shelf that I had nailed up in the woodshed. Finally, the tree swallows are starting to get busy at their nest box. The robin female built her nest in two days, but it will take the swallows as long as two weeks to finish their even simpler nests, probably because they are finicky and insist on using almost exclusively *white* feathers for their nest lining.

Every year a pair of northern orioles nests near our house in Vermont, and this year, 2008, when I watched our birds more closely than in other years, they returned in the first week of May. Seven days later, when the leaves on our black cherry tree were not yet full size, I was lucky to see the female pick a spot at the end of a hanging branch where she wove the first "threads" of her nest. I say threads deliberately, because that is what they look like—silver threads. I had always known that the baglike, silver-colored hanging nests of our orioles are made from fibers pulled off the previous year's dead milkweed stalks. But this time I went out to gather some of those milkweed threads myself.

No greens were sprouting yet, and I found many of last year's milkweed stems, dead though still standing, around our yard, but they had no frayed fibers hanging from them, none at all. I broke the brittle, hollow stems and tried to extract fibers. No luck. I could only peel off short, brittle fronds that had no resemblance to the trailing fibrous material (as much as a foot and a half long) that I had again and again seen the busy female bringing to the developing nest. She would wrap the thread around the slender cherry twig and then gradually weave it about and into other threads by poking her bill into and through the gradually forming fabric bag. She would pull each thread out where it poked through on the other side and pull it in again. After she had woven one of the long threads into the nest, she flew directly to the ground in a tangle of brush at the edge of our bog to get another fiber. Naturally, I went there too, to see what she was finding. Looking around carefully I at first failed to see any fibers, but there were old *fallen* stalks of milkweed, and from these, after I broke them, I could finally peel off bunches of long fibers. However, they were not exactly as silver as they seemed to be from my views of the nest through binoculars. They were gray and covered with a black coating, presumably dried, rotted tissue. Apparently she processed this raw material before it became useful thread for her nest-weaving.

Although the oriole female started building on May 12, she didn't finish the nest until eleven days later. The male never lifted a thread but sang busily nearby. I decided to examine the nest to be sure what it was made of. The climb of about fifteen meters to the nest in this cherry tree was arduous, but luckily for me I could pull the branch where the nest hung close enough to my perch on a limb so that I could take pictures. As expected from what I had seen during construction, the whole nest appeared to be made of the weathered but tough milkweed fibers. The bottom of the nest bag, however, was lined with totally different material: stiff, long brown fibers that were laid down in a loosely circular pattern. They looked like pine needles, only thinner and longer. Were they the locally superabundant dried narrow sedge leaves? No. I finally found that the best fit was the fiber that I found in dried dead bark of wild grapes. In between the brown nest-lining fibers was the main ingredient—a white fluffy material. A very abundant equally fluffy material was available on dried cat-tail stalks within a hundred meters of the nest, but the orioles had opted instead to use the fluff of now-maturing willow seed to line their nest.

Not all birds build nests. The early reptilelike ancestors of birds probably laid many eggs per clutch.[1] With the evolution of increasingly sophisticated care-giving, there were probably fewer eggs. When more than one egg was laid and the female stayed with them until the young, still helpless, hatched, the earliest nests may have been a mere hole in the ground or nearby material scraped around herself to form a ring to keep the eggs and young from rolling away. However, a bird that lays only one egg per clutch could potentially dispense with nest building, provided it has a relatively safe nesting spot where the egg would stay in place. For example, because murres lay only one egg per clutch (probably in part because they cannot rear two young at once), they are able to fit that egg onto a tiny space on a bare sea ledge. That is, the one-egg option may have opened up the possibility of nesting on inaccessible sea cliffs and thus gaining relative safety from ground predators. Similarly, the fairy tern lays only one egg, unlike other terns who usually lay two or three on the ground. It has the opportunity or possibility to lay its egg onto a fork on a branch in a tree. This option likely avoids the strong predator pressure on the ground. It also avoids predator pressure in the tree, since an incubating bird without a visible nest can totally hide its egg and would not be suspected of even being near an egg. The crested tree swift, which also has only one egg per clutch, has a minimal nest on a tree branch that barely cradles its single egg, and the bird strad-dles the nest to incubate it.[2] Emperor and king penguins are the only penguins that have single-egg clutches, and they also don't build nests because their feet hold their egg. This allows them to nest on the ice and tap into a rich food resource available in the open sea. One egg is all an emperor penguin *can* incubate, since there is nothing with which to build a nest on sheer ice. However, he (only the male incu-bates) places the egg on his feet to keep it off the ice, tucked into a downward-sloping belly brood pouch.

Two eggs would require at least a crude nest or receptacle. In many terns and shorebirds a mere scrape in the ground may suf-fice, and in pigeons and doves, who also lay only two eggs per clutch, the nest is a minimal affair, barely visible when the attending par-ent perches on it and incubates. Unless the nest is a scrape on the ground, few aspects of bird behavior seem more specific and perhaps idiosyncratic than nest building. From the placement of the nests to the materials used and the way they are assembled, nests are frozen behaviors that are, at least loosely, taxonomic markers and often indicate phylogenetic descent.[3] Thus, most titmice nest in tree holes or equivalent cavities within which they build fluffy nests of moss, hair, and feathers. Most swallows build nests of mud lined with fine grass and feathers. Finches build dense, compact nest cups in forks of tree limbs. Wrens build domed nests with a side entrance hole and they usually nest near the ground. Owls don't build nests at all, but use cavities or the nests of other birds, and shorebirds make crude scrapes on the ground. Woodpeckers hollow out solid trees and lay their eggs on the floor of the often deep cavity, directly onto the wood chips remaining inside. Geese and ducks make a nest out of whatever material they can reach from the nest site and pull it toward and under them with their bills. They then line it with their own feathers plucked out of their breasts. Many ploceid weaver birds form often ornate hanging nest bags from grass, but unlike those of the orioles, which are open at the top, weaver nests frequently have a long, downward-projecting entrance tube. The tailorbird stitches a single large leaf at the edges to create a bag. Swifts (Apodidae) don't gather nest material from the ground but use saliva to cement material such as feathers and aerial plant down that they catch "on the wing." They fashion feltlike bags or, in some species, make their nest entirely out of their own saliva. Nevertheless, the variety even within the swifts, weavers, flycatchers, swallows, and others is great, and therefore with much divergent and convergent evolution, nest variety is virtually endless and may seem overwhelmingly chaotic and idiosyncratic. The great crested flycatcher incorporates a shed snakeskin into its nest in a tree hole or bird box (although these days it often uses a piece of plastic wrap instead). The red-eyed vireo almost always garnishes the outside surface of its nest with wasp-nest paper. Some of these choices of nesting material are puzzling and they may or may not be adaptive. However, different components of the nest have obvious separate but nonexclusive functions, which include support, insulation, camouflage, and in some cases perhaps also medication, mate attraction, and the ability to repel predators.

*Support.* Most above-ground nests require a base that anchors the nest to the substrate. In herons, for example, that base consists of very long twigs and branches that bridge between often widely separated tree branches. In many weaver birds it may be a weave of grass fibers that anchors or ties the nest to a twig at the far end of a branch. In swifts it may be a cup of dried saliva attached to a cliff wall, or saliva used to glue twigs or feathers into or onto a vertical substrate like a cliff face. The mud constructions of a diversity of birds, including some swallows and phoebes, accomplish the same purpose. Whereas orioles use long, tough fibers to attach hanging bag nests onto the tips of branches high in trees, their relatives the red-winged blackbirds and common grackles commonly weave strands of grass and sedge between upright stems and cattail fronds, and the grackles additionally build a mud cup (that dries and becomes solid) on top of this framework. Although where I live in Vermont and Maine the orioles almost invariably weave their bag nests from the gray fibers of the previous year's milkweed stems, I found one nest that was made of another type of plant fiber, which I was unable to identify. Apparently their concept of "fiber" is flexible: I once found an oriole nest bag near a boat landing at a lake near Bishop, California that was woven almost entirely from pieces of light blue monofilament fishing line.

Using mud as mortar for nest building requires special techniques. A barn or a cliff swallow, for example, gathers a mouthful of mud to form a pellet, which is stuck onto wood or rock under an overhang. It must then wait for the pellet to dry and harden before it plasters on more wet pellets. It continues laying pellets like a mason lays bricks. Gradually, after several days, a saucerlike shelf (for barn swallows) or cavity (for cliff swallows and numerous other species) is formed. In some swallows the hard nest shell may be extended into a funnel, which forms an entrance tube. Then the nest lining of soft grass and feathers is inserted. If the nest lining can be considered as the nest, then making the clay receptacle is analogous to making a bird house. The woven nest of icterid orioles and many ploceid weaver birds is an analogous structure.[4] It requires fiber and weaving rather than mud and daubing, but it also creates a receptacle that serves as a foundation for a nest within it.

Rather than building space by daubing mud or weaving fiber to make an enclosure, birds also achieve the same thing by excavating sandbanks and termite mounds. Most famously, woodpeckers and some other birds excavate solid wood. Hole nesters and those nesting in burrows in the ground have built-in nest support, and indeed the hollow that they create functions in the same way as a bag nest for a weaver or an icterid. Some birds like woodpeckers use the wood chips that they don't remove from the bottom of their nesting cavity as a nest lining. Sand-tunnel nesters, such as some kingfishers, also do not make a proper nest lining—the sand suffices. However, many cavity nesters (parids, swallows, flycatchers) do make substantial nests inside the cavities and tunnels that they choose as their nest sites. Presumably that behavior reflects an evolutionary past before changing their choice of nest sites and taking along their habit of building nests in the more sheltered sites.

*Camouflage.* In those nests where concealment is important, the nest exterior doubles as camouflage or may incorporate specific material for concealment. For example, the lichens incorporated into a long-tailed titmouse nest built into the fork of a lichen-covered tree make it difficult to see the nest. A robin's untidy nest covering of trailing grass stems and loose leaves may be mistaken for a pile of debris. Eastern phoebes always garnish the outside of their nests with fresh green moss, which camouflaged when they nested on damp mossy cliffs in the northeastern forests but now makes them highly conspicuous at their new nest sites, on human-made buildings. A tailorbird's nest tucked into the fold of a leaf uses the leaf as nest camouflage.

*Insulation.* One important function of nests and nest lining is probably cushion and insulation for the incubating bird and the young, since the lining typically consists of soft, downy plant material and feathers. There is often a wide choice of materials, but in many species they are apparently not picked at random. Some birds exhibit seemingly idiosyncratic preferences. The dozens of tree swallow nests that I have examined in the bird boxes in Vermont and Maine have almost invariably been lined primarily with white contour feathers (although one outlier had no feathers) on a base of thin dried grass stems. An adjacent bird box with nesting house wrens is typically filled almost to the top with dry sticks by the male. If chosen by a female, it will be lined primarily with dark feathers. A third adjacent box used by a pair of European starlings was also lined with dark feathers and dry grass. Black-capped chickadee nests have a nest lining that is the consistency of loose felt, surrounded with a "filler" of fluffed-out green moss and animal hair. In a box adjacent to a tree swallow nest lined only with white feathers and one marbled wood duck feather, I found a great crested flycatcher nest that consisted mostly of last year's dried white pine needles. The eggs were cradled on a mixture consisting of a great horned owl feather, a number of Canada goose feathers, a dried snakeskin, and a tuft of raccoon fur. Meanwhile, the yellow warblers' nests in the nearby bog are lined with a solid white layer of willow seed fluff, and that of least flycatchers with tufts of pink (unidentified) plant down.

*Medicine and mate attractant.* There is a fairly large literature on various species of birds incorporating fresh green aromatic plants into their nests.[5] The green plants in starling nests are provided only by males, who bring them specifically in the presence of females and only during courting. After the females start laying eggs the males cease bringing in fresh herbs.[6] These observations suggest that these nest materials serve proximately as a sexual attractant. A male who goes to the trouble of collecting showy or scented nest material may offer a display and thereby exhibits "good genes." The offering of greens, however, may have direct advantages as well.

The greens that starlings use are generally aromatic, and since the plants have evolved the aromatic compounds as a defense against herbivores, they may then also protect their nestlings against parasites, pathogens, and blood-sucking mosquitoes.[7] In Philadelphia, for example, European starlings were found to prefer wild carrot leaves, which inhibited mites. Helga Gwinner and coworkers have followed up on the herbalist hypothesis in a systematic study with European starlings in Southern Germany.[8] Over a period of several years they removed natural starling nests from 148 nesting holes in their colony maintained along field edges in sixty bird boxes. They replaced half with artificial nests containing the herbs that were normally found in starling nests there and the other half with artificial nests containing green grass, which is normally *not* found in starling nests. They found that the number of mites, lice, and fleas were indistinguishable between herb and grass (control) nests. Nevertheless, nestlings from herb nests were heavier than those from control nests. Furthermore, as determined by blood parameters, these nestlings had healthier immune systems. The authors of the study speculated therefore that the herbs stimulated the immune system to better cope with the parasites. These were extraordinary results, and the main features of this study were justifiably repeated, this time including the time-course of the parasites in the nest as well as bacterial counts from samples of the nestlings' bellies. As before, the results confirmed (in each of five successive years) that the nests with herbs produce conspicuously larger (heavier) young.

*Innovation.* The amazing differences in the places and ways that different bird species build their nests appear to be a tribute to adaptation through the process of evolution. Nevertheless, behavioral programs are sometimes open to several possibilities. In New England common ravens use predominantly the knobby twigs that they break from poplar trees for making their nest platform, but in the far north of Alaska where there are no trees, a pair of ravens who made a nest platform in the girders of an oil storage tank used interlaced steel welding rods pilfered from a nearby construction site instead. The nest was "feathered" with work gloves, and Teflon and flagging tape. On the Great Plains a nest was found with its foundation built of snippets of barbed wire.

Individual variability in the face of a new environment is the raw material for the evolutionary process. And in general, conservative group characteristics are an indication of genetic inertia. All woodpeckers excavate nest holes and build no nest, laying their eggs on the wood chips remaining in their hole. Ducks and geese generally nest on the ground in marshes, wrens build closed nests having a small entrance hole, swallows usually use mud in their nests, and swifts commonly use saliva to glue nest material together. Wood warblers and flycatchers build cup nests on trees or on the ground, and kingfishers dig nest burrows in soil. Terns and shorebirds nest on the ground in mere nest scrapes. However, despite the general tendencies, there are almost always blatant exceptions that defy the "rules." There are species of ducks that are *restricted* to nesting in tree holes, a goose species that nests in fox and other ground burrows, swallows that dig tunnels in soil and nest in these burrows, a wood warbler, a wren, flycatchers, and kingfishers that nest in tree holes, a tern that lays its egg on a tree branch, a sandpiper that nests up in trees, and a murrulet that nests in the tallest redwood trees. And the birds that adopt these outlier behaviors do so not just occasionally—they do it "as a rule."

The distinct species differences that often depart far from the norm of the group are diverse among the titmice. Many species of titmice (Paridae) use existing cavities, primarily those made and used by woodpeckers in a previous nesting season, and then build their often substantial and bulky nest inside. Interestingly, some species are flexible in where they build. The great tit in Europe nests not only in natural tree cavities but sometimes also in discarded tin cans, shoes, mailboxes, and occasionally even mouseholes. I have often found the nests of the black-capped chickadee in my bird boxes in an open field, but I have also seen them excavate a nest hole in a sugar maple limb about seven and a half meters high. More commonly I find them excavating their nest holes from rotting gray birch stubs within a foot or two of the ground. Excavation involves both members of the pair, and it can take them two weeks to finish making just the place for their nest. The nest material of green moss and hair takes at least an additional week to accumulate. Left loose, it allows the female of these generally early nesters to cover her eggs when she leaves the nest. Only one brood is produced per year.

An even longer nest-building commitment is made by the long-tailed tit. It has evolved an innovation that substitutes using or making a tree cavity by weaving a bag with a tiny entrance hole at the top. The whole structure serves as a cavity. This tit also, like the others, lays up to a dozen white and red-brown spotted eggs per clutch in an enclosed space, but it creates that space by building, not excavating. These birds are among the most skilled nest builders. For a tiny bird of only seven to ten grams, its nest is huge—about twenty-three centimeters by thirteen centimeters. It is a heavily insulated bag with walls two centimeters thick made of moss and lichens like other tits, but held together with spiderwebs. It is fixed against a tree stem or crotch and camouflaged on the outside with bits of lichen from the immediate surroundings, making it practically invisible except for the small entrance hole near the top. It is like an artificial tree cavity created from the outside in, lined with up to 2,000 tiny feathers.[9]

Nuthatches, close relatives of the tits, are also tree-hole nesters, and they also build nests inside these and other enclosed cavities. However, their nesting behavior differs. The European nuthatch uses pre-existing holes, but they need not be perfect. If a cavity has cracks or irregularities, or if the entrance hole is too large, the female uses saliva mixed with clay to patch cracks and reduce the size of the entrance hole. If pine trees are available, she makes the nest exclusively of the scales of pine bark, but if no bark chips are available, she may use dry leaves instead. In contrast, in North America, the red-breasted nuthatch hammers out a nest hole from decay-softened wood and makes feltlike nests inside it like those of titmice. Since the red-breasted nuthatch makes its own nesting cavity, the entrance is the right size, but it plasters sticky pitch or resin around the nest entrance hole. Perhaps this material acts as a deterrent to pests or predators. I know of no other birds that plaster the nest entrance, except hornbills, who have independently evolved the behavior. But they have taken it to much greater lengths than any nuthatch.

Brown creepers, who are relatives of titmice and nuthatches, do not nest in tree holes as such. Instead they favor the space underneath loosely flaking sheets of bark. In New England I have seen their nests, usually under the flaking bark of balsam fir and sometimes of white pine. A sheet of bark that is attached at the top and flanges outward would appear to be oriented backward for installing a nest. To accomplish nest attachment, the bird first has to install a platform where there often appears to be no attachment point. But a combination of sticky spiderwebs and long thin twigs of spruce and fir (which have rough surfaces due to the nodes where needles had been attached) "grip" the trunk and the inside walls of the bark. The nest built on this platform is lined with soft plant down and fine strips of birch and cedar bark.

Making nest holes in sandbanks is apparently an adaptation for initially making a "foothold" to hold a nest, or nesting in holes or crevices that evolved by digging deeper to create a tunnel. For example, the European eagle owl occasionally chooses a nest site under sod which overhangs a sandbank, where a natural cavity is created. Such a situation suggests that a bird could evolve to scrape away some of the sand to make a better nest platform, dig a bit more to conceal the nest, and ultimately dig a burrow. Numerous birds have independently evolved to use or make such burrows for nesting, including some kingfishers (most of which use preexisting tree holes), mot-mots, bee eaters, and an owl—the burrowing owl. None of these isolated species nesting in sand burrows builds nests.

*Stereotyped behavioral rules.* A bird's nest is a physical record of a behavior pattern to produce a structure. The challenge is to deduce the specific series of rules of behavior that were required to shape it. Occasionally an unplanned "experiment" reveals something about what the bird "knew" or didn't know when it was building. Usually evolved responses are perfect or nearly so in that they account for the sum of experiences over evolutionary time, but when the animal encounters something it had not met before, then it may make mistakes. One thing birds have not normally encountered in their natural environment is a perfectly uniform nesting substrate.

A friend from Idaho recently told me about a pair of barn swallows that started to simultaneously build *two* nests side by side on a long straight beam under her porch. Then, curiously, the nest builders seemed to become confused and abandoned their breeding attempt when it came time to lay their eggs. Barn swallows normally come back every year to the barn or shed where they again use the same ready-made nest. As expected the next year a pair (unidentified) came to the same nest site but did not use either of these two pre-formed nests. Instead, the birds built a new nest directly straddling the tops of the two previous nests. Meanwhile, another pair of barn swallows had arrived, and this pair, as the pair had done the previous year, simultaneously built two nests side by side on another horizontal beam in a shed. They then laid four eggs into one of the nests and left the second one empty. (Curiously, two of the young "fell" out of the nest.)

The behavior of building two nests simultaneously side by side does not occur in the wild, and it suggests that something was amiss with the nest building on the beams. I had seen such behavior twice before under similar circumstances in two other species, once in phoebes nesting on a beam in our outhouse, and the other time in a pair of robins on a beam in our barn. Both pairs abandoned their nests after finishing them. In all three species there was one thing in common: all the nests were on smooth and perfectly horizontal beams, almost directly in between studs that were separated by nearly a meter.

Normally any one spot in nature where a bird might build its nest has something that distinguishes it from an adjacent spot. But in these four cases the birds were apparently confused about the precise spot where their nest was, or should be. They vacillated and two nests developed side by side. Difficulty came when the nests were finished and it was time to lay the eggs. The territorial species never have another nest like their own near them. Unlike often colonial swallows, they had never in their evolutionary history had to deal with another fresh nest "appearing" alongside their own construction. A situation like this normally never arises because under the natural conditions in the wild these birds build their nest on an irregularity of the substrate. Robins normally choose a place where a branch forks, or per-

haps the top of a stump next to an overhanging twig, and like phoebes, they tuck the nest into or onto a nook or cranny on a cliff. Both
the space and the attachment place for the nest are created by an irregularity, and the limiting feature is almost always space. Yet on a
horizontal beam of sawed lumber the bird may start the nest near
the middle, where the features are identical. Something in their behavioral program had misfired to produce behavior that in the normal context is not only adaptive but highly intricate. Now in a new
context it was bizarre and maladaptive. Are we in our newly created
world hugely different?

*Communal construction.* Birds that have a high social tolerance for one
another can exploit advantages of nesting next to each other. Colonial cliff swallows, for example, treat the walls of other nests much
like the walls separating apartments in a human complex. By building
nests adjacent to each other the swallows therefore save on additional
trips to carry in gobbets of mud, saving time and energy and speed-
ing up their nest construction.[10]

Monk's parrots do essentially the same but by using sticks.[11]
Parrots are largely cavity nesters, and the monk's parrot's ancestors
were presumably less fussy than other parrots about choosing a nest
site. They may have nested in clefts and crannies with large openings,
filled in with nest material to narrow the opening. Piling large sticks
together creates semicavities next to them, and so by adding new
nests onto existing ones a communal nest of many separate "apartments" can be created. Many weaver birds that don't weave nests,
including our common house sparrow, similarly sometimes switch
between using existing cavities to finding semi-cavities, which predis-
poses them to build free nests in or next to dense vegetation. The red-
backed weaver has gone a step further, building communal nests that
resemble those of the monk parrot.

The ultimate communal nest is that created by the sociable

weaver of southern Africa. The finished structure may have the size, shape, and appearance of a human's thatched hut. It may contain hundreds of "apartments" under one roof. Such structures can only come into and stay in existence in a very dry climate where what one generation builds is available for the next to build on; such structures may remain for centuries. The relative permanence of these communal nests has allowed these birds to adjust their lifestyles to the nests that they can live in year-round. They serve as refuges from the searing heat in the summer, reducing the birds' dependence on free water, which they would otherwise need for thermoregulation.[12] Similarly, in the winter the communal nest protects the birds from cold, reducing the amount of energy they need to spend.[13]

A golden plover nest and eggs. As is typical in most birds, the eggs of any one clutch may be irregularly spotted and blotched in the same way within the same clutch, although spotting and blotching may vary between species and sometimes within the species.

The coloration of a typical clutch of common raven eggs is, like those of most corvids, a blotchy greenish blue streaked and spotted with olive green and a bit of purplish blue.

This tree swallow nest contains a full clutch of five eggs plus two eggs dumped in by a black-capped chickadee. Both birds were nest-building and then laying in the same nest box, but the swallows "won" the nest occupancy. Or did they?

A nest and clutch of common grackle eggs. One egg is possibly a parasite dumped in by another female.

This nest of a sora rail probably contains two Virginia rail eggs. Can you find them?

A clutch of American coot eggs. Coots regularly dump eggs into other coots' nests. They discriminate against eggs that are not their own by shoving them to the periphery.

A clutch of sharp-shinned hawk eggs. As in many hawks, the color patterns commonly vary within the same clutch, with the first-laid often having the darkest pigmentation. Raptors likely do not have to discriminate against foreign eggs dumped into their nests because the adults are highly vigilant and fiercely territorial.

Ploceus capitalis
yellow-backed

Ploceus luteolus
Little weaver

Eggs = nat size

2 cm

Egg coloration in two African weaver birds, Ploceidae. Left shows the little weaver and four clutches of eggs. The birds nest alone, and the nest has a long entrance tube. The yellow-backed weaver nests in dense colonies, and it builds nests whose contents are readily accessible. The females lay clutches of two eggs each, each clutch quite different and specific to the individual bird. Twelve different clutches are shown here.

Puffin

Loon

Cormorant

Great Auk

a

b

c

Murre

d

e

2 cm

Nat. Size

Individual variation among common murre eggs *(center, center right)* is used for egg recognition. Great auks also had a single egg *(below left)* on sea ledges, and their eggs also had great color variation, presumably for egg recognition. Puffins, which nest in underground burrows, have white eggs *(above left),* and loons, which lay two eggs in a nest in the open, lay camouflaged eggs *(above center).* Cormorants identify their eggs *(above right)* by having separate nests.

A chipping sparrow nest parasitized by a cowbird egg. The cowbird female removed one chipping sparrow egg after inserting her own.

The same nest nine days later. The young cowbird demands most of the hosts' attention because of its big mouth, and it grows faster than the hosts' young. In this case the cowbird parasitism halved the sparrows' potential brood size.

A song sparrow nest with four sparrow and two cowbird eggs. One of the song sparrow eggs has a highly unusual spotting restricted to one end. I speculate it got stuck for awhile in the oviduct and then popped out quickly.

A northern oriole nest containing three oriole eggs and one brown-headed cowbird egg. Note the great variation of the fine and blotchy markings of the oriole eggs in contrast to those of the cowbird.

Unlike many other cuckoos, this North American species, the yellow-billed cuckoo, builds its own nests and cares for its own young. Only sometimes does it parasitize the nests of other birds, including those of its own species.

A tree swallow incubates in her nest box. Note the white feathers that arch over and partially cover her. These also shield the eggs when she leaves the nest.

A tree swallow nest with just-hatched young is lined with large white contour feathers. The eggs, which appear round in this end-on view, are pyriform.

The eggs at the bottom of the relatively dark interior of a marsh wren nest are also dark in color. Usually they are almost buried in a loose layer of brown cattail seed. In this case the nest was lined primarily with wood duck feathers.

In this section of marsh wren territory, at least eight "dummy" nests made by the male are visible.

Typical pink mouth linings as seen in most perching birds are framed by yellow-edged bills, as in these just-hatched cedar waxwings. The pink color is the blood seen through the thin skin.

The mouth lining of some parasitic species is color coded. *Above left:* the generic, most common color of songbirds. *Above middle:* the New Guinea waxbill *Erythrura trichroa.* *Above right:* Madagascar cuckoo *Coua ruficepts. Below left:* North American black-billed cuckoo. *Below middle:* African waxbill *Estrilda astrild. Below right:* Pin-tailed whydah, *Vidua macroura,* a nest parasite that mimics the estrildid finches' mouth markings, or vice versa.

9

## THE EGG

*I think that, if required on pain of death to name
instantly the most perfect thing in the universe, I
should risk my fate on a bird's egg.*

T. W. Higginson

ALL ANIMALS START LIFE AS A SINGLE
cell, an egg, but only birds have evolved to lavish so much energy on the egg itself. Birds stock that cell with massive amounts of nutrients so that it reaches enormous size. They enclose it in a hard shell colored with the most extraordinary variety and artistry and then release it into the world. Yet in most cases they are never far from it and treat it with incomparable care. Birds' eggs are a remarkable adaptive achievement because they have

permitted colonization from the coldest Antarctic ice into the hottest, driest deserts. Other animals may produce hundreds to millions of eggs that must be deposited into specific, generally moist environments, and they gamble that a tiny fraction will result in offspring. Birds' eggs, on the other hand, are so well designed and so well cared for that for some individuals as few as one or two are sufficient in a lifetime.

I have in front of me the egg of a Canada goose—the only one out of five seemingly identical ones that did not get smashed by a predator that attacked the nest in the bog by our house. It is colored uniform beige, and its smooth surface is a pleasure to touch. Relative to other biological objects it seems absolutely symmetrical. Though slightly pointed at both ends, one is more blunt than the other. If all eggs matched this shape then it would already be remarkable, but murres' and sandpipers' eggs are relatively sharply pointed at one end and much more blunt at the other, whereas owls' and other raptors' eggs are almost perfectly spherical. How can objects that are so variable with their precise curvatures yet uniformly perfect be shaped within something fleshy and flexible like the oviduct?

Birds' eggs, conservative in shape, are wild and variable in color as if they were painted by an artist who went to great lengths to create the most fantastic intricacies of lines, shades of color, squiggles, discrete spots, and cloudy blotches. The combination of the strictly conservative and the wildly radical make a contrast that is art of both Apollonian and Dionysian dimensions—truly art at its best.

*Egg formation.* Eggs are the key to a large part of the fascinating biology of birds, and they prompt us to ask how the egg is possible and then why this and not some other way? The answer to the first involves mechanisms. We have vague notions that the eggshell is formed for or by the specific egg contents because occasionally an egg contains a double yolk, and such eggs are much larger than the norm; and sometimes an egg is laid without a yolk, and such eggs are tiny. Only rarely is an egg misshapen.

Most birds lay only one egg a day during their precisely timed egg-laying stage, and they continue laying until their clutch is complete. Female birds have only one gonad, though males have two testes. The female's ovary is on the left side of the body and it looks like a bunch of tiny grapes. Each "grape" is an ovum that might leave the ovary to become an egg. Some birds, the determinate egg layers, produce a certain number for a clutch and then stop. Others, the indeterminate egg layers, such as chickens, keep on laying many more than their normal clutch if eggs are removed before the full clutch size is achieved. That is, some birds "count" out how many eggs they will lay before they begin, and others stop laying after a certain number are in their nest. However, except for very large birds, who have a much longer development period and hence more of a seasonal time constraint, most will lay a replacement clutch if the first clutch is destroyed.

An egg's yolk begins to accumulate in the ovary about a week before being laid. The yellow in the yolk is ultimately derived from plant carotenoids in the diet, and their concentrations may be regulated according to laying sequence and affect fledging success.[1] The germ layer that will become the embryo lies on top of the yolk, and when the yolk has achieved full size it is expelled from the ovary and fertilized with sperm stored from a previous copulation. Then it enters the wide, funnel-shaped mouth of the uterus. The egg develops further the next day as it travels down the muscular uterus. The uterine walls secrete additional nutrients, adding the albumen (egg white) around the yolk. The color of the yolk may vary from yellow to orange, depending on the amount of carotenoid pigments (directly or indirectly derived from plants) it contains, and laying sequence can affect the pigment concentrations.[2] As the growing egg travels farther, it becomes packaged in membranes and finally a hard shell of calcium carbonate. How is a layer of what soon becomes a solid and brittle shell formed? The materials for it are applied from a liquid solution, yet end up in a species-specific shape that does not warp during its passage through the oviduct. Finally, the egg is pushed out, blunt end first.

The eggshell is not just any solid, rocklike layer. The developing embryo will have to breathe, and the shell is equipped with thousands of tiny pores that are large enough for gas exchange yet small enough to greatly reduce water loss and entry of microorganisms. Bird eggs vary in size from that of a pea in hummingbirds to nearly the size of a soccer ball in the recently extinct elephant bird of Madagascar. Since gas exchange is by diffusion only, there is a great variation in the amount of gas exchange for incubation. Eggshell porosity and surface-to-volume ratio limit diffusion and hence could set limits to development of the embryo.[3] Larger eggs should have more and larger pores despite thicker shells. Larger eggs require longer for the embryo to develop, yet this lengthier incubation potentially poses a greater risk for contamination by bacteria. It is a mystery how the holes are made to just the right microscopic size through thick and thin shells and in just the right numbers.

The eggshells of most other animals are uncolored, including those of the reptiles from which birds evolved. There are some exceptions in insects. Birds' eggshells are unique in having often bright and intricate color patterns. The egg is "painted" on its white, light cream, or light blue background with pigments (porphyrins) that are derived from the hemoglobin of ruptured red blood cells. There are two kinds. One is bluish-green, and this pigment when present is distributed uniformly throughout the entire shell structure because it is deposited along with the calcium shell. The other pigment, which has a range of tints mixing red, yellow, brown, and black, is applied in spots, blotches, and squiggles onto the shell surface, although not necessarily until all of the shell is formed. These pigments collect in glands in the walls of the oviduct that have openings to the passage through which the egg is pushed. These glands have varying small and large apertures, through which their pigment is squeezed.

As the egg proceeds blunt end first, after the albumen, membranes, and shell have been added, the pigments ooze out and create their spots, streaks, and squiggles. Part of the color variety of these markings is achieved by the timing of the application. If the egg stays awhile and travels slowly, then the egg becomes painted with more and larger spots. If it speeds up, then there are fewer spots, and they appear streaked. The earliest markings, before the full complement of shell has been laid down, are paler and milky; pigments applied along with the shell and hence covered with a thin, translucent layer will be filtered and muted by the shell's background color. Those same pigments applied after the shell has been laid down tend to be bright and bold.

The color pattern is achieved as if innumerable brushes hold still while the whole canvas moves. If the egg remains still, spots result. If it moves, there are streaks, and if it turns as it moves, then there are lateral stripes. This is probably not the whole story, however. Any one of numerous specific markings (as in icterids and murres) are scribed in several directions at once, even when they are very near each other. Do muscles in the oviduct twitch and move locally in different directions to make the color-releasing glands move like a brush on a stationary canvas?

*Incubation.* Aside from the coloring and shape, both critical for the safety of the egg, there is also the direct behavioral intervention of the adults that serves as the equivalent of "mothering." Eggs' contents remain untainted by bacteria for many days, but microorganisms eventually enter through the pores through which the developing embryo also receives oxygen and gets rid of carbon dioxide. The membranes around the egg inside the shell presumably provide an added if not the main physical barrier to bacteria while permitting the diffusion of gases in and out. It is essential for the egg surface to remain dry to discourage the entry of bacteria, yet the relative humidity around the egg must be high enough so that it does not dry out.

Incubation helps to achieve both processes, yet in many birds it does not begin until a week or two after the egg is laid—until the clutch is complete. There is, however, also a great potential cost to delay, especially in a warm and moist environment where eggs could spoil through infection. Is this one of the reasons why tropical birds have fewer eggs than northern birds?

The fresh egg must also solve another physiological problem. Many birds lay such a large clutch that eggs may be laid as much as two weeks apart. But they must all hatch at the same time. There must be a way to stop growth of the embryo completely and then start it again. Growth cannot occur at a slower rate at a lower temperature, as is the rule for most biochemical reactions. Growth must commence full-tilt precisely at the beginning of incubation, or else the eggs will hatch at different times. For example, suppose a duck lays a clutch of a dozen eggs, laying one egg each day. If she started to incubate after laying the first egg, then there would be almost a two-week span between hatching of the first and last egg. In most birds with precocial young, the entire clutch hatches nearly at once, and all the young leave the nest hours after the first egg hatches. Any late-hatching young would be left behind. Air temperatures prior to incubation, when the eggs are laid, may be high, in which case development must not be just slowed but *stopped,* or it may be very cold (such as in winter-nesting birds), and in that case the eggs must be protected from freezing. There are therefore likely biochemical mechanisms that involve "measuring" the egg temperature to regulate embryo growth at a near-constant 40 °C. It is also likely that behavioral mechanisms of the parents respond to their monitoring of egg temperature. Either or both of the parents sit on the eggs to warm them passively, of course, but incubation also involves the parent's morphology and physiology, and to some extent vocal communication during the later stages of incubation with the young, who peep when they are cold, which induces the parents to incubate.

Birds do not have major feather tracts on the chest and belly; the bare skin in these areas is exposed to the eggs to provide direct contact during incubation by spreading the feathers of the lateral feather tracts apart. In many species this alone is not enough. As the time of incubation approaches, most female birds develop a brood patch. The skin of the area that will contact the eggs becomes heavily vascularized with blood vessels. It becomes thick and turgid and resembles a giant blister. Males, who may also take their turn on the eggs, don't have a brood patch. The brood patch apparently has sensors for keeping track of the egg temperatures directly, and the bird can then assess when it needs to incubate and for how long. The softness and flexibility of the vascularized skin permits more contact with the eggs, and as the blood is flushed through it, heat flows into the eggs by conduction. Additionally, incubating birds usually have warm feet (to help heat the eggs). When they are not incubating, they keep their feet near air temperatures or just above freezing at very low air temperatures to conserve body heat, using counter-current heat exchangers of their blood circulation that allow for an uninterrupted blood flow to the legs and feet.

There are two main considerations for care of the eggs: safety from predators and proper temperature to ensure development of the embryos. With a premium on fast development came the necessity of warmth such as an incubating bird with its eggs in a snug nest can provide. But not all birds lay their eggs cushioned in a soft nest. Some birds lay them instead on bare rock where a thick eggshell compensates for the lack of a nest.[4] The bird's ancestors probably accomplished both protection and perhaps also temperature management by depositing them in hidden places that were warm, as most reptiles do now. Burying them in decaying vegetation that provides heat from decomposition, either scraped together or brought in, might have been a further refinement. Crocodilians still do this, and it is also practiced now in the megapodes, large turkeylike birds of Australasia. Possibly two separate routes of evolution then developed—one to tend the mounds, to carefully regulate their temperature by opening and closing them, testing the temperature and then adjusting the compost. The other technique is to sit on the eggs directly and incubate them. This process requires maintaining close physical contact as well as periodically turning the eggs, perhaps not only to heat them evenly but also possibly to prevent membranes from growing together and sticking to the shell.

The demands and durations of incubation, even with eggs maintained fairly steadily at the same temperature, vary considerably. Important variables include egg size, nest construction, ambient temperature, and development rate achieved at hatching time. The small eggs of most perching birds hatch in about ten to fifteen days, and those of swans and eagles and ostrich may take thirty-six to forty-five days. Although the nest as such probably doesn't affect the number of days until hatching, it does affect the effort required to get there. Most birds, especially tropical ones, do not incubate continuously except at low temperatures. At high air temperatures they may take frequent breaks and leave the eggs unattended, while at low-temperature incubation breaks can only be brief. Ducks and geese pluck the down from their belly feathers and then cover their eggs with these feathers when they leave for brief incubation breaks to forage. In a sense they leave a part of themselves behind to continue keeping the eggs warm. Grebes have found another way to keep their eggs warm. They cover their eggs with additional plant material on their floating nest of decaying vegetation when they leave. In this case the function is also likely to hide the otherwise conspicuous white eggs and to reduce heat loss from them, but the heat generated by decomposition would also warm the eggs.

The Canada goose egg that I was fondling had been the only one left intact in the remains of a nest where, in the struggle with a coyote or raccoon, the goose had been dragged off and her eggs scattered. The one egg that remained in the nest in the afternoon was covered with down. I had not seen the goose on the nest all day and realized only in the afternoon what had happened. When I checked on the nest that one egg, thanks to its "blanket" of down, still felt warm enough for me to attempt to "incubate" it with a heat lamp. Unfortunately it overheated to 50 °C, at which point I knew it was dead. Bird embryos are usually not harmed by cooling, but they are easily killed by heat. Indeed, birds attending their nest are careful to prevent overheating as well as cooling. Shorebirds may stand over their nest to shade the eggs in intense sun and also to wet the eggs after soaking their feathers to cool them by evaporation.

In some birds, particularly those species in which the female is cryptically colored and the male is brightly colored, only the female incubates. In others the sexes take turns incubating, though not necessarily in equal portions. In mourning doves, for example, the male incubates without interruption all day long, and the female takes over in the evening and incubates until the next morning. As already indicated, there are also special cases where the female leaves the eggs soon after laying them, and the male then takes full responsibility for the entire incubation and the caring for the young. However, even in many cases where the female does all the incubating, the male's responsibility is not less because he brings in the food to fuel the female. Female crows and goldfinches, for example, do all the incubating, and they make specific food calls when hungry. When the male feeds her, she acts like a baby bird, and he feeds her in the same way as he feeds the young.

*Hatching.* The thickness of the eggshell is a compromise—thin enough to conserve calcium, which is often a limiting resource for egg production, and thick enough to reduce breakage when the parent sits on it and rolls it. Tampering with this fine balance can have disastrous results. The pesticide DDT, used widely to control mosquitoes and other insects, caused eggshell thinning that resulted in the near extinction of peregrines and other raptors. In some birds, such as penguins nesting on rocks, the eggshell needs to be thicker than in others, in whom it is cushioned in a soft nest. Lastly, the shell must be thin enough for the chick to be able to escape. The eggshell can be a prison, and the parents provide no help for the young to escape it. Why not? I suspect it is a much too risky strategy for programming parents to be able to open the egg at just the right moment, and so escape from the shell has been left to the occupants. However, they are given a tool for that escape.

Baby birds before they hatch have soft bills, but they have a pointed, sharp projection on the top of their upper mandible or bill, the "egg tooth." It works like the tip of a can opener. The young bird also has specially developed neck muscles to move its head back and forth, and by scraping the top of the bill with the egg tooth on the eggshell it eventually wears a groove and then a small hole through it. By hatching time, the embryo has absorbed some of the calcium from the eggshell, so the shell is not as hard as it was when first laid. With a puncture in the shell, air can come in, and the bird, increasing its metabolic rate, can work harder and exert more physical effort to enlarge the hole. After an arduous struggle that may require over a day, the hatchling slips out through a hole in the side of the shell. But in a clutch of more than one egg, the hatchling does not emerge alone. The final synchronization of emergence involves vocal communication among the young before they hatch. In precocial birds like quail, who leave the nest shortly after they dry off, the young peep within the shell and thereby communicate their hatching readiness.

## 10

### PARENTING

### IN PAIRS

*I now believe that mothering is so critical and so challenging that to force anyone into its service is immoral.*

Anna Quindlen

JUNE 2, 2007. THE PILEATED WOOD-peckers that nested in the hole the pair excavated in a poplar tree near our house had recently been silent (after being noisy in early spring). But yesterday morning I again heard their repeated bouts of drumming, the usual woodpecker song. I suspect their young just hatched. This morning at 8:30 AM the phoebe also made a huge commotion. But it was not the agitated exclamations one hears when a chipmunk or a cowbird is near. Instead, I heard their cheerful "phe-be" endlessly and emphatically repeated. "The young have hatched!" I thought, and so I immediately went to the nest in our shed, reached up to poke my finger in, and as expected felt the soft bodies of a couple of hatchlings as well as a couple of still unhatched eggs. There were no empty eggshells either in or below the nest, so the parents must have carried the shells away.

*July 13, 2005.* On this gorgeous morning our house wren is gathering feathers to line her nest in their little nest box at our window. This will be their second brood this year. Alder flycatchers and mourning doves are also singing again and getting ready for their second (or third) broods. Many first broods were destroyed this year because of the rains, and our first brood of phoebes was raided by chipmunks. The pair then built a new nest in the woodshed, but the extended rain kept the insects low, and one by one starving young fell or crawled from the nest. Several days of sunshine at the right time make a big difference in whether or not they can raise a second brood. Some species never try. The tree swallows and the red-winged blackbirds never raise a second brood of young if the first is successful, which is puzzling because they are the earliest nesters. And even after their young are fledged they both disappear and to join wandering flocks of hundreds or thousands.

*June 8, 2007.* I awoke at dawn at my camp in Maine hearing the drumming of yellow-bellied sapsuckers coming from all directions. They seem to find the most resonating sounding board, which in this case happens to be the tin pipe chimney to the sauna on my sugaring shack. Their forceful drumming was joined by a red-eyed vireo's singing in the sugar maple grove next to the cabin, and the phoebes, who had a nest with just-hatched eggs on the beam above my bedroom window, added a few refrains. It was time to get up, and I staggered groggily to start a fire in the iron stove, brewed some coffee, and then went to check on an ancient sugar maple where I had weeks earlier heard sapsuckers hammering out a nest-hole in a dry stub that was slanting at 45 degrees. I had before seen them nest only in vertical poplars, and they almost always hammer their nest holes into the live poplar trees that have a specific fungus (*Fomes igniarius* var. *populinis*) growing on them. The presence of the fungus signifies a softened core. I had never before seen a woodpecker make a nest hole in a dead branch like this one that slanted so sharply. The nest entrance was on the lower side of the dead branch. This pair had "gone out on a limb" in more than one way, and I wondered how they were doing.

They had met disaster. We had just had a very strong windstorm, and their branch had broken through where they had hollowed it out, crashed to the ground, and shattered. Five dead naked young sapsuckers lay spilled on the ground. I had never before seen such strange-looking baby birds. Their lower bills were about 10 percent longer than their upper, and they all had a thick callus on each heel. The latter would be useful for birds like them who spend weeks growing up jammed onto a wooden floor with no nest to cushion them. Having that callus when they must stay vertical on a narrowly cramped hard floor seemed like a viable substitute for a nest. They did not have a shred of down on them, unlike most altricial birds, who have at least a few light feathery plumes. Their bellies bulged huge and black out of their thin white almost transparent skins, revealing their internal organs in sharp detail. I could see bright pink lungs and winding intestines, but most prominent were their large stomachs, bloated with ink-black contents.

Not wanting to let the sapsuckers' misfortune go to waste, I did an autopsy—at least 95 percent of the black matter in the sapsuckers' stomachs was half-digested black ants. The stomach contents of one "typical" nestling weighed 2.9 grams, and the "empty" bird itself weighed 4.5 grams; it had swallowed 64 percent of its body weight in ants, which would be the equivalent of me, at 160 pounds, eating 412 quarter-pound hamburgers without the trimmings at one sitting and being ready for more (as I presume these young birds were at the time the tree limb crashed). If I had had four youngsters eating as much, I doubt that I could have provided them with 1,248 hamburgers every day, but then, mercifully, if I had been a woodpecker I would have had to provide for my kids for only a little over a month, not twenty years.

Parenting is a full-time occupation for most of the summer in those birds that come here to breed. It involves long preparation, an intense period of provisioning, and a continually shifting panorama of critical decisions. One of the first and perhaps most critical parenting considerations of a female bird or the pair is the number of eggs to lay before completing the clutch. Birds must also consider how many clutches to risk having in a season. Evolutionary logic predicts that birds produce the most young that they can over their lifetime. But this often requires restraints, because laying and incubating too many eggs can sometimes result in starving all the young if there is not enough food. Clutch size and timing have to be adjusted so that they incorporate a safety factor that takes the typically changing and often unpredictable vacillations of the weather and food supply into account. On the other hand, some birds routinely create more offspring than they can raise, which results in siblicide and sometimes also cannibalism.[1] As Douglas W. Mock and colleagues have shown, the apparent "extras" are back-ups that act as an insurance policy in case one or more of the eggs don't hatch or something happens to some of the young.[2] When to begin incubation affects the outcome of the competition among siblings, because if incubation begins before all the eggs are laid, then the early-hatching young become the favored "core" offspring and the later-hatched will be the "marginal" ones because of their smaller size. Those that hatch later are then at a competitive disadvantage—they often show reduced growth and higher mortality. Thus, maternal decisions allocate the outcome of family strife, and the marginal offspring pay the price of the parent's initial over-production of eggs.

*Scheduling.* The reproductive "decisions" cannot be left to conscious logic. They evolved through experience of the species and some of them are genetically controlled and tied to endogenous schedules keyed to the seasons. Perhaps some of the best studies of the complex interactions of environmental versus endogenous programs that affect the scheduling of reproduction are those on the stonechats. In the wild, Siberian populations (long-distance migrants) of this bird typically lay six or seven eggs per clutch but have only one clutch per year. The central European populations (short-distance migrants) lay five eggs and are multi-clutched, whereas those near the equator in Kenya (nonmigrants) lay only three eggs per clutch and are single-clutched in the field. Under natural conditions the chats native to Africa restrict themselves to about three young per year, Siberian birds to six or seven, and the Europeans about fifteen (in three or more clutches). Are these responses due to direct effects of the environment or to genetic programming in response to the environment? To find out, Eberhard Gwinner, Sibylle Koenig, and Chris S. Haley at the Max-Planck Institute for Behavioral Physiology in Germany have reared many hundreds of stonechat clutches in aviaries.[3] They held pairs over successive breeding cycles under indoor versus outdoor conditions and under changing European photoperiods or constant photoperiods of near 12.25 hours light and 11.75 hours of darkness. They found that clutch size is largely determined by the genetics of the female, not by the male, and only to a limited extent influenced by photoperiod. Barbara Helm and colleagues from the same institute have continued these studies.[4] They found that Siberian chats still reared only one clutch per year, and under the same constant conditions birds from the temperate and equatorial populations stayed in reproductive condition and produced two or three clutches in a year. A Siberian female chat paired with an equatorial male (who will sing and be in reproductive condition) did not develop ovaries after her short nesting time window was past. Similarly, when a Siberian male is paired with an equatorial female she did develop eggs, but he would not mate with her after his short breeding season. In other words, environmental cues are overridden by endogenous breeding schedules (which are also tied in with endogenous molting and migration schedules that vary among birds from one area to the next).

In contrast, pigeons and doves, and swifts and hummingbirds, regardless of where they live, typically lay only two and sometimes only one egg per clutch, although the number of clutches per year varies from one to seven depending on local breeding seasons.

Timing of available food is probably the main driving force for the evolution of breeding schedules and egg number, but ability to provide enough food for the offspring is not the only constraint that determines their number. Increased nest visits due to many young could increase predation pressure, and in addition the increased workload of providing for many young can increase stress.[5] In birds of paradise, for example, where the males take no part in nest building, incubation, or feeding of nestlings, a typical clutch is two eggs in the lowlands and only one egg in the highlands (where less food is available).[6]

Rearing many offspring per clutch can reduce longevity and compromise future reproduction. In one study involving 200 pairs of kestrels in the Netherlands, the brood size was manipulated by transferring two nestlings from one nest to another. The parents that had raised two extra nestlings had a 60 percent greater chance of dying a half-year later than did those with the clutch of two fewer nestlings.[7] Similar delayed costs due to parenting stress have also been reported in rooks and collared flycatchers, and males and females may invest differently in response to brood manipulation.[8]

*Destroying signs.* Incubation is, for the most part, a time of pause after the often demanding nest-building and egg-laying intervals. But immediately after the eggs hatch and before the babies are fed their first meal, a bird's heavy work of parenting starts anew. As soon as the young hatch, the first task of the parents is to dispose of the eggshells, and near hatching time in the spring I commonly see the bright blue of a discarded robin's eggshell on the driveway, and have found a fresh shell of a raven's black-blotched olive egg a half-mile from the nest.

Nikolaas Tinbergen, a Dutch behavioral ecology pioneer who received a Nobel Prize for his work, was one of the first to wonder why birds seem to have an apparent strong aversion to empty eggshells near their nests.[9] He speculated that the newly exposed white surfaces inside the shell make the nest easier to detect by predators, and he put his hypothesis to the test with experiments involving predation by gulls and crows. He placed normally camouflaged black-headed gull eggs and white-painted eggs into the field and learned that the white eggs were taken more frequently. He then placed empty eggshells near the hidden khaki gulls' eggs and found that the presence of empty eggshells made the hidden eggs more likely to be found, thus eggshell removal (and camouflaging egg color) has functional significance. However, there is a cost to removing eggshells. The parent has to leave the nest, and a predator could then rush in and raid the nest of the newly hatched chicks. Tinbergen observed in his gull colony that even neighbors sometimes rushed in and ate chicks in the nest before they had dried off if the parents left them to remove eggshells. Colonial gulls were slower to remove their eggshells than solitary nesters such as oystercatchers and ringed plovers. Since gulls are not averse to eating their neighbors' eggs and freshly hatched chicks, Tinbergen concluded that it pays the gulls to have at least one of the pair stay at the nest until the young can hide, and hesitation to remove their eggshells could reflect the fact that predation from colony mates is greater than that from crows.

*Benign neglect and consequences.* Most birds must contend with a fluctuating food supply. A time of great plenty may be followed with a time of scarcity as fruits, seeds, insects, and other prey are pulsed through time. Present food supply is often a relatively good predictor of future supply, or at least that often short period during which they must provision their young, and some birds gauge future food needs by adjusting clutch size. For example, during an outbreak of lemmings in the Arctic in spring there is a relatively good assurance that there will be much food after the snowy owls' young hatch, and so they adjust their clutch size accordingly, laying ten to eleven eggs rather than not breeding or laying only two to three eggs per clutch in times when lemmings are scarce.[10] But there is a downside to this strategy: if they lay too few eggs and a food bonanza comes later, then they lose out on a reproductive opportunity. They hedge their bets by starting their incubation after the first egg is laid (while the male continues to bring food) so that if the food supply crashes, then at least the first-laid eggs yield young that survive and only the later-hatched young are sacrificed. If they laid too many eggs early and started incubating only after all are laid and then the predicted food supply diminishes, they might lose their whole clutch to starvation.

Food caching, such as practiced in a number of corvid birds, can be one solution to a fluctuating food supply by providing some predictability or help in an emergency. But meat-eating raptorial birds have an additional option: their own young can become an emergency food store when starvation would otherwise threaten the whole clutch. Thus for them laying extra eggs is not necessarily wasted reproductive effort but instead a modest investment as well as an insurance policy.

If laying "extra" eggs is an adaptation and not just a mistake, then one would predict it would be found primarily in birds who consume starved siblings or are so much larger than their nestmates that they can overpower the smaller and either evict them or eat them.

This requires asynchrony in egg hatching, so that the elder ones get a head start. Cannibalism of live young by the parent is, I suspect, not an option, because parents must and do have a strong inhibition to killing at or near the nest; they are selected for tolerant behavior. They instead may have a number of adaptive reasons to produce a surplus egg or surplus young as they gamble on the future: 1) to have a replacement egg if one doesn't hatch; 2) to track an unpredictable food supply; 3) to replace low-quality offspring; and 4) to facilitate siblicide in case of food shortage.

To find out if asynchronous hatching is an adaptation largely of parental behavior, it is instructive to compare such birds with those that have to do the opposite. In ducks, geese, shorebirds, and grouse, for example, good parenting requires ensuring that all of the eggs hatch almost at the same time so the young can leave the nest together. Those that don't hatch in time get left behind. Synchrony is achieved by the parent not starting to incubate until all the eggs of the clutch are laid. In addition, peeping of the chicks when they are nearly ready to hatch helps synchronize their emergence.

In a number of seabirds, herons, owls, and many eagles and vultures, asynchronous hatching is a prerequisite of a different parenting strategy, one in which large demands for food provisioning coming from a fluctuating supply cannot be sustained. Eagles, for example, can seldom raise more than one chick at a time, although they routinely lay two eggs per clutch. Some eagles practice "obligate brood reduction"; the older sibling kills the younger even if there is plenty of food in the nest. Also, having just one egg is potentially risky since if it fails to hatch, the entire breeding attempt fails. Large birds such as eagles and vultures do not have time for a second breeding attempt that year. The bearded vulture in Spain, for example, has a breeding cycle requiring six months of offspring care. Both parents are normally required to provision for one hatchling, and they are seldom if ever able to feed two. The vultures' second egg is laid a whole week after the first, hence the unusually long delay between eggs appears to be a mechanism that facilitates siblicide.[11]

Herons typically lay five eggs in temperate regions (two in the tropics), and although all of the eggs normally hatch, sometimes only one or two chicks leave the nest. Similar to eagles, owls, and many hawks and falcons, herons start to incubate as soon as the first egg is laid. As a result, the successively laid eggs hatch in the order in which they were laid. One that hatches days after the firstborn faces a huge handicap; in times of food scarcity the firstborn, who is stronger, outcompetes its siblings and gets most of the food. Heron young do not eat each other, but they can and often do commandeer the incoming food by shoving the competition out of the nest. Egret nestlings often engage in bloody battles against each other using their beaks as spears.[12] The parents do not interfere. This is, for herons, good parenting; if all were fed equally, then all could perish during times of food scarcity.

In the vocal intraclutch competition to be fed among young herons in a nest, there is as in other birds selective pressure for them to beg loudly, since that behavior stimulates the parent's feeding response. The less the young are fed and the more there are of them, the louder they scream for food. However, such noise attracts predators. It is important, therefore, that they have safe nest sites. Generally only those in safe nest sites (such as ravens on cliffs or woodpeckers in tree holes) can afford to be noisy.

All of a number of great blue heron colonies that I have seen in Vermont were in beaver bogs in tall, bare trees standing in water. Typically, as beavers dam a watercourse a several-acre pond results that kills the trees (usually pines), which the beavers do not cut down. Then the herons come and build their big, bulky nests high on the dead, bare branches. Their choice of trees standing in water presumably serves to deter some tree-climbing predators who could hardly

fail to see the very conspicuous nests and hear the noisy young. However, as I learned at a heron rookery, there are no guarantees.

It was in the first week of May 2008. The maples, cherries, and
birches had just put on their delicate yellowish-green leaves, and with
them had also emerged the swarms of blackflies. The first vireos,
warblers, and rose-breasted grosbeaks were returning. Great blue
herons nest early, and I presumed they would by now have young. I
loaded my kayak onto my truck and drove to a friend's beaver bog
where there had been fifteen to eighteen nests in previous years. I
floundered through the woods to the heron colony, carrying and then
dragging my kayak. To my great disappointment, when I finally arrived at the heronry there were, this year, only two nests. One heron
stood still on one nest at the tip top of an almost-bare pine tree, and
another flew over the pond on heavy wingbeats to land on a dead
snag. I settled myself under a pine tree near the shore and watched
and waited. A half-hour later, neither heron had budged; I was not going to see much "parenting" here and I wondered what had happened
to wipe out the colony.

Joanne Ignewski, living adjacent to the bog, filled me in. She
had watched "her birds" for almost three decades and declared that
the experience of watching them over this time had shown her that
"Nature is crueler than I had ever imagined." Like woodpeckers or
ravens, safe in their nests, the young herons had called loudly for food.
In effect, the young do not only endanger themselves, they also black-
mail their parents because the noise they make until their stomachs
are stuffed potentially attracts predators that could destroy the whole
brood.

Ignewski observed the young squawking in the evening as the
sun set when the parents flew in with food. Raccoons then came and
swam out to a nest tree and easily climbed it. Then she heard the piti-
able crying of the young as the raccoon dragged them off into the

woods. "Night after night the raccoons came, and they were not de-terred by water." She contacted the Audubon Society for help, who "did nothing—they just told me they would not 'disturb the natural system.'" To her credit, Joanne did disturb the natural system. She did the natural thing (but too late) of nailing metal flashing around the bases of some of the trees with nests—the same kind of metal flash-ing that state fish and game departments put around the bases of poles that hold up the bird boxes they put out into bogs and marshes to provide breeding places for the wood ducks that may then be hunted. But it was too late. The herons abandoned the colony.

The example of herons highlights the general problem of lay-ing too many versus too few eggs. Although the parents might "want" to lay as many eggs as possible to have many young, they pay a cost if the competition among the young then causes them to scream so loudly for food that they give the nest location away to predators. Both the number of nest visits and the volume and number of beg-ging cries are the main cues predators use to find nests, and most bird parents have vocal signals that cause their young to become si-lent when danger threatens. Bird parents determine how often and how much food to deliver by the begging calls of the young, who compete among themselves by begging loudly (and ultimately also by evolving colorful, attention-getting mouth colors). Raven young, who are usually safe on cliff sites, can sometimes be heard screaming from a kilometer or more away, and woodpecker young safely inside solid tree holes beg almost nonstop and can be heard from at least a hundred meters, yet I have never heard the begging calls of a ground-nesting bird, which are very brief and soft. Thus, there has been selec-tive pressure for reticence in begging, for reducing its volume, and for discrimination in calling; young songbirds beg at almost any dis-turbance, whether it is a vibration or noise produced by a parent or predator, but as they grow older they become increasingly specific and beg only in response to stimuli from their parents.

*Predator diversion.* I jogged down our road to the Huntington River for a swim by a gravel bar that is partially overgrown with vegetation and heard a spotted sandpiper call, but when I looked I did not see it nor did one fly away. Somehow the bird had vanished. I immediately suspected that it had settled onto a nest, because a sandpiper had nested last year on the same gravel bar. I waded across to it and almost immediately saw the bird fluttering ahead of me as though it had a broken wing. As on several previous occasions when I had surprised a mallard hen as she was leading her ducklings to water, the sandpiper repeatedly fluttered directly in front of me, offering herself as a target. To most young and naïve predators, such a display is realistic enough to induce them to make a quick dash and attempt a pounce, and so they are diverted or lured from the nest or the young crouched motionless on the ground. The broken-wing display is common among diverse groups of ground-nesting birds. It is practiced not only by shorebirds and ducks but also by some sparrows, warblers, finches, grouse, doves, and owls. It has therefore evolved independently on numerous occasions. However, on this occasion the "predator" found the nest because of and not in spite of the broken-wing display.

The nest was a mere mold in the gravel and partially covered with grass. It held one still-wet chick with matted down. Another was already dried, gray and fluffy, and it got up onto its spindly tan legs to run off among the loose stones, where it hunkered down. The nest still held two tan black- and brown-blotched eggs that were cracking, and I saw the bills with the egg teeth of the emerging chicks. Only one large, empty eggshell was conspicuously lying next to the nest, so the second one had perhaps just been carried away when I arrived. Thus, the sandpiper had not only provided me with false evidence by faking a wing injury, it had also been in the process of removing real evidence of its young.

Small solitary-nesting songbirds can also provide effective predator diversion by false signaling. I had noticed that chickadees may try. The local black-capped chickadees commonly nest in holes they excavate in rotten birch stumps, and when I tapped at one nest in such a stump where the bird was incubating, I was surprised that it did not flush. Instead it made a hissing-grunting noise from inside the hole. As long as she didn't show herself, that sound represented something I couldn't even imagine, and almost all animals shy away from the unknown. Burrowing owls nesting in ground-squirrel burrows mimic a rattlesnake's rattle, and that tactic may at times be backed up by the real thing; their burrows commonly do contain snakes. The chickadee's display was pure bluff.

Small ground-nesting birds face quite a challenge when beleaguered by a potential predator. Their first defense is to try to be neither seen nor smelled, and for that the nests and the incubating birds have to be camouflaged. Some, such as woodcock, may be distasteful, although this is pure speculation. Most bird dogs, I've been told by a bird-hunting friend, are very reluctant to pick up woodcocks though they eagerly grasp ducks and grouse. Fluffing out and bluffing to try to look big when you are as small as a sparrow is not likely to repel even a weasel. But faking injury may cause the same animal to try to attack even a duck or a grouse.

Many parents of precocial young in the face of danger give specific calls that induce their young to scatter, hide, and "freeze." Ruffed grouse often make a screeching sound as they flutter in front of you and land twenty to thirty feet ahead to then run on the ground with their head down low. Mourning doves, who nest in trees and cover their young with their bodies until they can fly, flutter to the ground and act as though they are unable to fly if they are forced to leave when you climb the tree and come too close to the nest.

We tend to think of the bird's nest defenses as stereotyped behaviors that are executed mindlessly. But a varied strategy is best if not also necessary with predators that learn from experience, like I do sometimes. The research of Caroline A. Ristau with plovers shows that their actions are flexible and allow them to behave in ways that manipulate their adversaries perhaps intentionally.[13] For example, plovers who initiate the broken-wing display to a fox will fly into the face of a cow that walks toward the nest until it changes direction. At Barrow, Alaska, I once examined the nest of a semipalmated sandpiper several times in succession. I first found the nest only when the bird flushed after I had almost stepped on it, and it fluttered off in the broken-wing display. The next time I came near the same nest the bird also allowed me to get close, but this time it scuttled off realistically mimicking a running lemming or a mouse. When I came again on another day the bird flushed from a distance and called loudly from nearby. Many predators, in this case probably Arctic foxes, would not be fooled by the same ruse several times in succession.

*Bodyguards.* Protecting the young is one of the more important parenting functions. Given the advantage of size and weapons, it can be exerted through physical confrontation. Large birds of prey are pre-adapted for the role. Peregrine falcons stoop on prey larger that themselves and hit them in midair at some 200 kilometers per hour, knocking them out and then swooping back to pick them out of the air. Even goshawks, who hunt mainly grouse and hares, are intimidating at their nest. Usually they at least do not attack frontally as one watches them, but the mate will then attack from behind the observer. All the while the birds keep up a loud and strident staccato clanging. The display is intimidating because it is backed up with long sharp talons and the talent to use them. The closer one gets to the pair's nest, the louder the birds become, and the more frequent and closer their swoops onto the intruder. When they come so close that you feel the rush of air from their wings and feel that they would next rake your head or back, it is time to back down and retreat. Great horned owls, eagles, and red-tailed hawks have also bloodied those attempting to reach their nests. Female golden eagles are used by falconers on horseback in Mongolia, who chase wolves and then re-lease their bird onto the fleeing wolf. The eagle strikes it from behind and clamps its mouth shut with their talons. The hunter can then ap-proach and dispatch the prey. In raptors, unlike in geese, it is the fe-male who is larger and better able to be the defender of the nest and young. The size difference as such is thought to be derived from a segregation of diet, so that the smaller males catch the agile prey and the females catch the larger prey that is more difficult to overpower. Together the pair thus avoid hunting competition near the nest.

Swans, who may weigh about as much or more than some large eagles, and some geese who equal or exceed eagles in size also are big enough to put up an effective defense, even without strong beaks and talons. Instead, they can deliver stunning blows with their wings. Although geese are close relatives to ducks, there is little com-parison between them when it comes to parenting. A successfully parenting female duck has no use for a male trying to defend her from a coyote or raccoon—her mate would only draw them in and so be a handicap. A male who stays with his mate will only endanger his own young, so he should stay clear of her and their babies. She can use concealment as a defense, and if that fails, she can feign injury. A gan-der, on the other hand, can put up a credible fight, and the one thing he will *not* ever do is feign injury. Instead he is much more effective exaggerating his strength, and that he will do well.

Goose and swan ganders accompany their mate and the gos-lings continuously, and are a deterrent to primarily mammalian pred-ators such as raccoons, skunks, coyotes, and weasels. In the popula-tion that I studied in Vermont, as soon as the young can travel (about two or three days from birth) the goose parents—with the female in the lead and the male taking up the defensive rear—always leave the small breeding pond area and enter the forest. They walk several kilo-meters to another area that is rich in browse and has open vistas. In this feeding and growing area, in contrast to the breeding and nesting area, several families join together to form large crèches of young.[14] That is, some of the parents who bring their clutch of young end up with two or more clutches of them, while others "lose" or give up their young. The adopters can accept new young presumably because it has little cost: all the young geese feed themselves on the plentifully available greens and there is no competition among them for food. Furthermore, it is advantageous to do so since, unlike in altricial birds where each new young is another mouth to feed, having others' goslings is sharing of risk; if a fox, hawk, or other predator does manage to get past the parental guard, then the more young the parents have adopted, the lower is the risk that the one taken will be their own. Indeed, in ostriches (where one cock has several hens) several females may lay into the same nest, and all the young then are united into one group. In those cases in which the young need to be fed, however (such as in flamingos and pelicans), where the young also crèche into sometimes huge aggregations, they are still fed by their own parents. The communal behavior is not altruism to save the others' lives—it is a shelter for their own young.

Confrontation in nest defense is used only when evasion fails, as was evident from the wild Canada geese I've watched and inter-acted with for about ten years in the beaver bog next to our house in Vermont. Before Canada geese finally nested there in 1998, I had occa-sionally seen a pair stop by. One year I saw a pair that was constantly together, and I was hopeful that they might nest. One day when I went down to watch them I saw only one—and it had its head low and swam into a cattail thicket. That behavior meant it had some-thing to hide; a nest was near! I started training my binoculars on a small peninsula in the pond where a series of sedge hummocks could potentially provide a nesting site. Sure enough, a fleck of white moved, and I then saw movement behind a veil of dry sedge leaves. A goose. She lay on her belly, and eventually she betrayed her pres-ence by pulling something toward her—she was making her nest.

The gander was now nowhere in sight. I climbed down from the maple tree where I had had a panoramic view of the bog, and I then waded across the marsh. As I came closer to the nest, she put her head down low as though to hide. Her white face patch would have been visible otherwise. I looked at her peripherally to disguise my intentions of coming nearer. However, when I was about four meters from her I stopped and then looked directly at her. In that moment of eye contact she jumped off the nest and started calling. Her mate then rushed out of hiding and came at me, honking wildly. They both increased their clamor as I neared the nest.

The nest with its precious cargo of four greenish-yellow eggs was surrounded by dry brown sedge and wet green moss next to dark water. Delicate shoots of light green marsh grass poked an inch or two above matted vegetation. It was an entrancing view, and I could have stayed many minutes, but I talked softly to the geese, and after I had now let them know that I knew about the nest by pointedly looking directly into it, I slowly backed off, continuing to look at them and talking softly to them as I left. I had been impressed by the changing tactics of nest defense as much as by the beauty of the nest, its setting, and the eggs, and since surreptitious behavior would alarm them, I talked to reassure them that I was not going to harm the nest.

Almost a month later, on May 27, 1998, I went again to the nest, now for the third time, to see if the eggs had hatched. As the goose skipped off the nest this time, I was surprised to see three of the cutest yellow-down puffballs imaginable, plus some eggshells that were squashed down into the nest material, plus one still-intact egg. This time she called only briefly. The gander came, but having recognized me and perhaps now knowing I was harmless, they both had become downright tame and tolerant of me. Normally birds get increasingly defensive of their nests and young the more they have invested in them. To test whether their tolerance was indeed due to recognition and learning, I went to a neighboring beaver pond to meet another pair at their nest that I had never visited. Here the gander flew straight at me and genuinely intimidated me as his wing brushed violently across my head when he flew directly at me. Although I was then still three meters from the nest, I did not care to come closer.

Smaller birds can gain considerable defense by communal effort. I have been intimidated (on the Isles of Shoals, off the coast of New Hampshire and Maine) by black-backed and herring gulls. At one nest that was relatively isolated from the others of the colony, where my daughter handled a chick, I came back to take pictures and was twice shat upon and once knocked in the back of the head by one of the attending pair. They took turns diving at me from behind. I would not dare to enter a colony of hundreds without wearing a hard hat. The red-winged blackbirds and common grackles in the bog near my house do not intimidate me, but as mentioned earlier they do dissuade any crow or raven (and perhaps cowbird) that comes near when they gather in a black swarm and attack.

Parenting is a job often involving more than just the parents. Sometimes the kids help out, especially in birds like crows. The off-spring of one or several previous nesting seasons may stay around before they themselves nest and possibly inherit the parents' territory. Although there is debate about whether they actually help much, the devotion of both parents as well as helpers to the current young can inspire our admiration. A friend, Albert Reingewirtz, told me the following story of how he came to his love of crows. He had been living and working in Eilat at the edge of the Negev desert when one windy day he saw something he said he would never forget and which, he said, "caused me to have a yearning I almost fulfilled." He had seen crows flying up and down in a big swirl of wind where dust and a newspaper were being sucked high up into the air. Crows waiting at the top would catch the paper and dive down with it. Then they would release it, and the wind would take it up again, and another crow would repeat the process. On seeing this joyful play, his yearn-ing was to have a crow as a friend. Later, his wish almost came true when a neighbor, knowing of it, brought him a young, barely feathered-out crow freshly out of the nest. It was being carried in a box, but the crow's parents nevertheless followed this neighbor the whole way. After seeing the parents' devotion to their young, my friend that night decided to return the stolen baby to them the next morning when he found them still around his house. He opened the box and grabbed the young crow to throw it onto the roof. It rasped in fright and anger, and "instantly around ten crows came flying from all around for protection." After that, every time that he came out of the house the crows would come and scream at him. He experi-mented by having his wife leave the house while he went in, and al-ways there was then silence "as though a switch had been turned off." When he went back out "all hell broke loose again." He concluded: "I don't know how to say 'I love you' in crow, but I now do understand 'I hate you' in crow."

*Feeding.* Possibly the biggest challenge most birds with altricial young face as parents is feeding their young. Feeding up to a half-dozen or more youngsters growing to adult size in about a week to ten days requires constant parental effort. I counted the number of nest visits the phoebe parents made to their nest on the window ledge of my cabin: twenty-eight per half-hour one late afternoon and forty-three per half-hour in the morning, which extrapolates to about 1,700 for-aging trips by the pair in a twelve-hour day.[15]

In a radically different parenting strategy of many species, the parents lead their young to the vicinity of food and the young then feed themselves. This parenting, involving precocial young, is found in Gallinaceous birds, shorebirds, rails, geese, and ducks, and others where the chicks leave the nest within hours or a day of hatching.

One or both parents may lead the young. The parents also warn them of danger, but then the young usually employ their own anti-predator behavior in conjunction with the parents'.

"Precocial" refers not just to physical development but also to behavior. I sat next to a nest of Canada geese, watching the goslings several hours after they hatched and had just dried their yellow fluffy down. The gander perched nearby on a beaver lodge, and I was in my kayak lodged against it and next to the nest. Goslings peeked out from under their mother's belly feathers, and when she stood up I saw them stretch their wing stubs as though they were fully-formed wings. These downy young preened their wing stubs using all the correct motions despite there being no sign of any flight feathers. They pecked at everything in sight, and in a few trials they distinguished food from nonfood. They ran, swam on top of the water by paddling, then even dove under the water. They flapped their little wing stubs while stretching tall in the same motion as the adults in what Konrad Lorenz called the "triumph" ceremony ("satisfaction gesture" is I think more accurate because triumph implies conflict and the gesture is in no way restricted to post-conflict situations). Two of them would later recognize one human from another (and follow him). They had a repertoire of very specific calls given in appropriate circumstances that even the human caretaker could interpret. Forty days later the goslings were much bigger, but they were still mostly covered in down. Contour feathers were coming in on their wing scapulas and secondary feathers began to cover their backs (presumably acting as rain guards). Their wing primaries, the feathers required for flight, were only starting to show as pinfeathers. Although "precocial" refers to behavioral development, it is not related to independence from parents. Swans, geese, and cranes may lead their flightless young around for months. Geese and cranes are physically and behaviorally very precocious, but they are dependent on their parents even into the adult stage in the fall, when they follow them on and are thus taught the migratory route to their winter quarters.

To appreciate a goose's precociousness one must compare it with, say, a baby warbler that comes out of the egg pink, naked, and blind. Six days later it is feathered out, and in two more days may fledge and fly. In a few more it becomes independent, but for the first few days of life its behavioral repertoire was limited to responding indiscriminately to almost any mechanical stimulus, such as one's moving finger, by raising its head straight up and opening its mouth, expecting food. At this stage it will swallow whatever drops into its maw. It has one other behavioral program: after being fed it raises its hind end to the nest edge and presents a fecal pellet in a most conspicuous way to its parent, who eats it or carries it away.

In most grouse and pheasants, only the females are involved in parenting. They lead the young and show them food, but in rails and cranes both parents lead the young and continue to bring food to them and hold it out to them so that they take it from the parents' bills. The young of others, like geese, find and identify food on their own. In turkeys, chickens, and quail, parents may pick at food and the young then follow their example and so learn to find food on their own. Loons and grebes carry their young chicks on their backs some of the time, and a parent may initially bring food to the young and hold it out to them to take. This requires bi-parental care, for even when the young are older one parent forages while the second serves as a mobile floating nest. The parents of many precocial birds also shield the young from rain and cold by covering them with their bodies, using their wings as a rain shield as the young huddle underneath. Woodcocks may pick up one or more of their young in or on their feet (personal observation) and then fly off with them, carrying them away from danger.

The great diversity of parenting behavior and dependence

on parents in precocial species is also reflected in the evolution of flight capability of the young. Alternately or in addition to defense by the adults, the young may quickly grow a special set of small wing feathers. In the megapodes of Australasia, who are "extremely preco-cial," the young not only fly within a day of hatching, they never see their parents and are totally independent of them from the moment they hatch. Similarly, in the ruffed grouse that I am familiar with in the Maine woods, the chicks are flight-capable within several days of hatching. By contrast, in the chicks of large birds such as geese and swans, which are protected from predators by the male parent and live in relative safety in or near the water, the chicks do not prioritize the growth of wing feathers by having an early temporary set (like we have milk teeth). Instead, they *delay* the growth of wing feathers for a couple of months, and when these feathers do grow in they are already the large and eventual final set used on the fall migration.

A great diversity of birds can be characterized in the intermedi-ate category between precocial and altricial. Gull, tern, and flamingo babies are down-covered and leave the nest after several days. Bit-terns and owls may also leave the nest long before they can fly. Great horned owl young, for example, are covered in white down when they hatch, but their eyes remain shut for another eight to ten days. They then shed their first down and grow a second coat. They are un-able to thermoregulate and their mother broods them most of the time for another two weeks while they acquire only about 25 percent of their capacity to regulate their body temperature. The male, who has fed his mate throughout the egg-laying, incubation, and brooding period, also sometimes stockpiles large amounts of prey in the nest before the eggs hatch, possibly in response to the young, who vocal-ize a day or two before they hatch. After they hatch and while the fe-male broods them, he still brings all of the food and gives it to the mate, who pulls it apart and feeds bits of boneless meat to the chicks. The young leave the nest when they weigh about 75 percent of their adult weight. The male may then feed them whole mice and other large prey that the nearly-grown young swallow whole. They can't yet fly, but they are expert climbers. They call for food incessantly and rely on their parents for food for months afterward. Hawks, falcons, and eagles are also covered with down at an early age, but they don't leave the nest until they can fly. None of these birds gape. In most raptors both parents hunt, and when the young are small the male commonly brings whole prey animals to the nest, and the female then tears off one bit at a time, holding it in front of the chicks, who take it from her bill. Later, when the young can tear the prey apart themselves, the adults merely drop it onto the nest.

*Extreme parenting.* In both precocial and altricial birds the juvenile stage is the most vulnerable in the bird's life, and there has likely been great selective pressure to reduce its duration. The many small passerines nesting on the ground or other vulnerable places do not have the privilege of a long, protected childhood. Half will on average become prey before they get the chance to fly well. The amount and kinds of food they consume have a great effect on how long they remain juveniles. Baby birds are made of protein, and protein is the main raw material required to rapidly build up their bodies. Most songbirds are therefore able to achieve phenomenal growth rates in their young by feeding them an almost pure diet of insects, although as adults their diet may consist largely of fruit.

But both the duration of incubation and the length of the fledgling period are a function of body mass, and thus size itself is a large part of the equation. For example, eagles, vultures, and swans incubate forty-three, fifty-one, and forty days respectively, whereas after hatching they require 70, 100, and 125 days until they are capable of flight. In contrast, in most of the small perching birds, the total time from fresh egg to flight may be as short as twenty days. In most the incubation as well as fledgling durations are each about twelve to fifteen days, and those who diverge from that do so for adaptive reasons; the European cuckoo, for example, has at twelve days a slightly faster egg-hatching rate than its hosts, suggesting an increased selective pressure to hatch ahead of its parasitized victims so it will be able to evict them from the nest. Similarly, ground-nesting passerines become flight-ready in a week to ten days after hatching, whereas woodpeckers, parrots, and other birds nesting in the relative safety of tree holes do not fly out of the nest until they are thirty to thirty-five days old.

The eggs and young of cave dwellers nesting on high rock walls are also relatively safe from predators, and they can afford to develop slowly, which means that they have the option of a less nutritious diet. Perhaps one of the best examples is the oilbird or guacharo of Central and South America. The guacharo is a frugivore (and a caprimulgid or nighthawk relative who is the only member of that group that is *not* insectivorous) who raises its young so deep in dark caves that it echolocates to negotiate in and through them. The oilbird's safe nesting sites have freed it from an insectivorous diet because the young can remain in a helpless stage for months. The bird's two young have an extraordinarily slow growth rate, not leaving the nest until they are 110 to 120 days old, when they weigh 600 grams, almost 200 grams more than the adults.[16] The grotesque obesity of the young is thought to be the result of having to eat enormous amounts high-caloric fruits in order to extract enough protein. However, fruit are easy to "catch" and their excess weight should not greatly hamper their foraging.

Swifts (Apodiformes) provide a unique parenting situation. Like the oilbird, they also nest in relatively safe nest sites such as caves and cavities, and they also have greatly extended nestling durations (to about forty-five days). However, they are aerial plankton feeders and have evolved to become the most completely aerial bird in the world. Arguably, the best-known member of this group is the common or European swift. The British ornithologist David Lack provided nesting places for them so he could watch their home life from up close, immortalizing them in his 1956 book *Swifts in the Tower.*

The common swift was originally at home nesting in clefts and crannies on Alpine cliffs, and later it switched to pseudo-cliffs such as the old stonework of medieval castles (hence its German name *Mauersegler,* or "stonework swift"). It has now become a common summer resident in many towns all over Europe, and it nests in small colonies in nooks and spaces under the roofs of houses (and occasionally in bird boxes). Now, as houses are built tighter there are fewer nesting places for them, and there is growing public concern for these fascinating town residents. The German Society for the Swift has taken it upon itself to encourage architects and builders to deliberately install clefts suitable for swift nesting. In the extraordinarily hot summer of 2003, many swifts were driven by the heat from their nests under rooftops, and the society oversaw a record 522 swifts installed in private homes to take care of the injured birds.[17]

The common (European) swift has, like all the other eighty-five to ninety species of swifts, achieved the ultimate perfection of life on the wing in the open skies. The two young from a clutch of this small (forty to forty-five gram) bird with long narrow pointed wings spend about six weeks in the nest and leave it at about fifteen minutes after sundown, apparently to avoid falcons. From that moment on the young's ties with their parents are broken forever, and they ascend into the sky to 2,000 to 3,000 m in altitude to meet up with others, staying continuously and uninterruptedly airborne for two to three years.[18] They do not touch down to Earth until after they return to Europe to nest, after having by then migrated twice to Africa. (Juveniles may sometimes, under emergencies of low temperature and lack of food, roost in foliage.[19])

For a long time the swift's aerial existence far from human eyes was doubted. Don't birds have to roost to sleep? It had seemed like a fairytale that they might sleep on the wing, but for once a fairy-tale turned out to be true. World War I pilots in open cockpits flying on moonlit nights were the first to report seeing the swifts drift by at 3,000 m (almost two miles high). Radio-telemetry data later confirmed that the birds ascend to such altitudes on their first evening when they leave the nest.[20]

As might be expected of a bird that is ready to fly at speeds near 200 kilometers per hour on the day it leaves the nest, swifts have an extraordinarily long nest residency—at least six weeks versus six to ten days for most songbirds of similar size. Day and night, but especially in early morning and evening, for weeks before fledging the swiftlets prepare for flight by preening their feathers and doing flight gymnastics in the nest. Both at night and in the daytime one can hear their excited voices and fluttering as they try their wings in the nest.

The swifts' diet consists of often tiny insect plankton such as winged aphids, and the parents bring their young compacted hazelnut-sized packets containing each 200 to 1,500 tiny insects that are stuck together with saliva. But insects don't fly on cold rainy days, and such weather may persist for weeks. In order to eat, the swifts have to follow the weather. When the weather is unsuitable for flying insects the parents (who can survive without food for only about four days) may fly thousands of miles to find better weather and bring food back to their young. However, unlike the young of songbirds, the swiftlets can survive starvation and wait out the weather by going into a hibernation-like torpor for about two weeks. That is, the parents may appear to desert their young for a week or more at a time, but the young, though seemingly dead, are not forgotten and will revive when the weather improves and the parents return.

Similarly, I suspect that unreliability of a continuous and predictable food supply may also result in temporary suspension of offspring care and account for the extremely long and variable incubation times of fulmars (forty-eight to fifty-seven days), gannets (forty-four to 130 days), and vultures (fifty-one to one hundred days). However, the aerial insect predators, the swifts, need to be expert flyers in order to be able to catch any food at all, and their long juvenile period in the nest, despite their strictly animal diet, is the added time needed for them to assume life on the wing. Their safe nest sites make a long juvenile development time possible, which in turn affects their future lifestyle. Indeed, time and energy constraints may also shape breeding systems.

*Nest hygiene.* The one parenting duty that precocial birds do not need to worry about but that is an issue for most altricial birds is cleaning up the poop from their young. There are prodigious amounts of it. Most altricial perching birds package their feces into membranous sacks that the parents eat when the young are still small. The fledglings' digestion is not fully developed, and there is still some nutrition to be had from the feces. When the young's digestive capacity is better developed the parents carry the fecal packages away to drop them at a distance from the nest. Hornbills regurgitate the solid, indigestible parts of fruit in pellets (much like those of bones and fur famously vomited by owls and other raptors, and ravens and many other birds that eat food with hard parts). These pellets are left in the nest and absorbed in the grass and leaves that had lined the nest. In addition, insects then colonize the nest and eat these wastes. I once examined the nest hole right after a pair of pileated woodpeckers had fledged their young, and the bottom of the nest hole held a deep layer of finely macerated insect exoskeletons and perhaps thousands of small fly larvae mixed in wood chips. Nest sanitation always has a cost, and some birds can't afford it.

I was very surprised when I closely watched the breeding cycles of a pair of mourning doves nesting in an apple tree next to our house. They let the feces accumulate in a ring all around the nest. (Ring doves in a laboratory colony had been reported to keep their nests "free of fecal matter throughout the course of a reproductive cycle."[21]) The birds must have kept their flights to and from their nest to an absolute minimum, or have done it when I was asleep, because I never saw it. I therefore presumed that their behavior made sense for predator avoidance—throughout the day they never once left either eggs or young (even large young) uncovered.

Marsh wrens nest over water, and although the parents carry their feces out of their nest when the young are weak and small, the young later back up to the nest entrance—the small lateral hole in their globular nest—and defecate through the hole and out of the nest. In woodpeckers and swifts, who both require over a month to fledge their young, both parents help keep the nest clean for about three weeks after the young hatch, but then after the young start to feather out they relax sanitation.[22] In woodpeckers enough waste matter accumulates that any unhatched eggs may be buried. The young of most species with cup nests may be able to keep the nest clean themselves, but the parents of most species remove every drop of feces as soon as the young defecate; no feces accumulate under, on, or by the nest. Hornbills, in whom the female is entombed in a tree hole by having the nest entrance cemented shut to leave only a slit, solve the nest hygiene problem by projectile squirting of very thin liquid, aimed to fly out through the slit. Raptor young have similar very liquid feces ("mutes") and powerful gut muscles for projectile squirting their mutes over the nest edge, or in some cases, such as hole nesters, aiming them up through the entrance hole of their nest cavity. As a young boy I once raised a pair of kestrels—a common small hole–nesting falcon. I kept the young in a cardboard box next to my bed in my room, and was soon surprised to find that they, unlike my young crows, squirted their mutes clear over the top of the box and far away onto the floor in the direction of the window.

Nest hygiene may appear to be a strictly parental duty, but it is a highly evolved behavior in which the main burden can be either on the parents or the young, or on both in a close cooperation. Even more amazing to me than my kestrel was a baby phoebe that I once had. I got it from our neighbor, who had parked her car directly under the nest, and her cat found the car to be a perfect stepladder to reach the nest and kill one of the parents. It rained for several days, and the one remaining parent could not keep up with both supplying the young with food and sitting on them to keep them warm, too. Three of the still-naked young, whose eyes were still closed, were soon dead, and I took the remaining one, barely able to move. It revived on a heating pad, and after being heated to about 29 °C it moved feebly. It could raise its head and open its bright pink mouth, making barely audible high-pitched cheeps, but only for a second. It begged briefly whenever its makeshift nest was jiggled. It was a challenge to get a morsel of crushed insect into its gullet in that narrow time frame, but I understood the briefness of its beg to be an adaptation to minimize alerting predators. Normally when the parent comes to feed the four or five young, it does not hesitate but drops the food it brings into the first mouth it encounters, and so if a baby is not fed instantly, then it will not be fed at all, except possibly at the next visit, which could be a minute or two later.

When I did manage to drop food into the baby phoebe's mouth, it swallowed and then, as in a reflex, instantly turned around and within a second presented its rear end to me, voiding a fecal pellet, before immediately dropping back down to sleep. It did not do this at every feeding, but it did not do it in between feedings. Thus, the parent has only to wait a second at the nest, and in that time, while already there, it attends to nest hygiene. Now I knew how the phoebes could keep their nest so immaculately clean without much extra effort. The young gape reflexively to the parents landing at the nest edge, and the parents are there to intercept the fecal sac the second their bill is free from the food delivery. The whole transaction of food delivery and waste disposal is an enviable example of parent-offspring cooperation. However, as I will next explain, although most passerine birds do practice such nest hygiene, some species, including ravens, barn swallows, and even phoebes, are at times very messy and let their feces drop directly under the nest.

*Starling parenting.* In the 1890s about a hundred European starlings were introduced to New York's Central Park. The species has since multiplied to a couple hundred million and spread to occupy the whole North American continent from the Atlantic to the Pacific and from Mexico to northern Canada. It is generally regarded as a pest, although it has been the object of many rewarding studies, and many people (including Mozart) have found them to be delightful pets. The husband-wife team of Meredith West and Andrew King wrote about the joys of their singing and mimicry and shed light on Mozart's fascination with them.[23] I have always enjoyed having these colorful, glistening, iridescent birds with their yellow bills and pink legs (in the breeding season) near me, and in May 2008, a pair nested in a bird box out of a hollow cedar log that I had hung on a post by our garden. Four young hatched out of the four pale blue eggs, and having just discovered some wonders of bird poop, I decided they were a gift from heaven and so decided to make some observations and take notes.

Hole nesters may have less to worry about from predators than young in open cup nests, and so they may not need to get rid of all feces whose scent could attract predators. How would common starlings nesting in a bird box handle their wastes? On June 3, 2008, the soil below the nest box with the clutch of baby starlings on the post at the edge of our garden had remained clean since the four young starlings hatched fifteen days earlier. I lifted off the top of the box on several occasions and there were never saw any messes inside, either. Thus, the parents must have been meticulously removing their babies' wastes. Then, on the sixteenth day (about three days after the babies were clothed in feathers), huge gobs of bird droppings suddenly appeared in front of their box—twelve, to be exact. I had counted them because as I had learned some years earlier and was still learning from constant reinforcement, parenting may and usually does involve not just feeding the young but also cleaning up their messes, and I wanted to know the magnitude of this parenting task, having had personal experience. But the birds have it easier. At first their feces appear to be plastic-wrapped, but later as the young of some species get older, the tight little gobs become huge runny splats instead. These of the starlings were now tending toward the latter size and consistency.

My starlings were presumably within days of fledging. Possibly the parents were no longer cleaning up after them because these young were now capable of keeping their room clean themselves. I lifted off the box cover and looked inside: still no sign of feces. All the baby birds were immaculately clean with not even a tiny white smudge on any of them. Perhaps the parents had either eaten or carried off all of the fecal material before this, and now the babies, or the parents, were dropping it in front of the door.

I covered the telltale sign on the soil at the edge of our garden in front of the box with fresh soil so that I could keep track of anticipated future deposits. I didn't have long to wait. The next day by evening there were thirty droppings, and in the next four days there were fifty-two, eighty-two, seventy-six, and eighty-six, respectively. With so much out-going, there had to be a lot of in-coming, so I was curious to find out how often the two parents brought food.

I got up at dawn to watch and see what was happening. At 5:30 AM the ground was clean (I had made a count by flashlight at 10:00 PM the night before). I brought along my binoculars, pencil and paper, and a strong cup of coffee, and sat down on a big rock at a comfortable watching distance of a hundred feet away. Within a couple of minutes one and sometimes two of the young starlings peered out of the oblong entrance hole to the nest at the same time. These young seemed extraordinarily alert, agitated, and curious. With quick jerky motions they cocked their heads all around—looking up, down, to the sides. Sometimes one or another leaned so far out of the nest box that I thought it would launch itself and fly off. Then I would hear the long drawn-out rasping contact call of an adult coming with food, and the babies would reply with higher-pitched calls. A second later the adult—beautifully attired and glistening in a green, purple, and blue sheen—would land on the top of the box as a dull-brown young-ster raised its bright yellow gaping mouth upward. In one quick, deft motion the adult shoved an insect down its gullet.

In the first hour the parents made thirteen visits, each time bringing in a total of at least nineteen (all dead) grasshoppers and one huge dead caterpillar, whose entrails were spilling out. This caterpil-lar was sparsely covered in bristles apparently because most had been pulled or knocked off. Extrapolated for the fourteen hours of day length that the parents feed their young, they make about 170 feeding trips per day. All of the starling's foraging flights were in the direction of a distant field, and there would have been no grasshoppers in be-tween. The field was at least a half a kilometer away, thus they were daily flying at least a total of 170 km, or on average 85 km each, just to be able to feed the young. I presumed that house cleaning involves multitasking and maybe some cooperation from the kids.

In the late morning, at 8:45 AM, after taking a break because I was tiring just watching them, I made another one-hour watch and observed that when I thought a bird brought in a grasshopper it was often two or three at once, because while it was poking one into a youngster's gullet another one or two sometimes fell to the side and down to the ground—which the parent would chase down and imme-diately retrieve. Curiously, although I could clearly see the in-coming, I didn't see much out-going. Most of the time one or two young birds were blocking the nest entrance; I saw no bird sticking its derriere out the hole between them. The adults' visits were brief, and neither of them ever entered the nest hole to possibly retrieve wastes. Only once did I see one of the adults perch at the entrance, reach its head in, and grab a fecal pellet. It then dropped it after several meters of flight. It could only have reached the pellet if the young had helped by perhaps acrobatically extending its hind end upward, almost to the nest hole. On the other hand, on several occasions I saw fecal material "fly out the door." Apparently some of the young were now projectile defe- cating. I concluded that the adults at this nest had stopped most of their nest hygiene duties, and the young were taking over and eject- ing their own feces. The switch in nest hygiene mechanism—from parental to self-propelled—appeared to coincide with the time that the young were first able to appear at the nest entrance.

The parents never went inside the nest hole, and these rela- tively grown young were fed strictly according to the queue that they themselves established by waiting at the nest entrance. Presumably the hungriest will be the most eager and the pushiest. Does any other have a chance? I expected them to leave the nest very soon, so I felt I had a very limited time to conduct an experiment to find out if a bird that is not in the queue at the nest entrance would still get fed. So I got the stepladder, climbed up, and lifted off the nest-box top. Reach- ing in, I lifted one of the young out of the nest and let it fly in our garden. (The sides are surrounded by wire screening to keep out large rodents and deer.) The bird flew directly into a patch of raspberry bushes and hid there.

My third hour watch was from noon to 1:00 PM. As soon as I got there I realized that my turd-counting was done; there was silence at the nest; the rest of the young had fledged. I had apparently just barely missed seeing them leave, and I regretted not having stayed continuously to find out if and or how the parents coaxed them out. However, I decided to sit still and watch anyway.

One of the parents arrived in twenty minutes. It had no food in its bill but made rasping contact calls. Immediately the left-behind youngster that I had prematurely fledged answered from the raspberry patch. The adult flew off rather than toward it, but returned fifteen minutes later with a big grasshopper in its bill, called again, and then flew against the outside of the wire garden screen from where the baby had called; it was trying to get in. Finally it flew over the top of the wire enclosing the garden and went directly to its hidden young. It most probably fed it the grasshopper because it was empty-billed when it came back out of the raspberry patch several seconds later.

Meanwhile, I located the rest of the clutch in a pine tree 200 meters from the nest box. The three remaining young were hopping nervously in the dense conifer foliage as I got near. If this family behaved like most starling families, then in a few days it would be traveling around together, one parent with some young and the other with the rest of the brood. I was now free to examine the nest. There was no sign of feces in it at all. It contained some old dry grass and mostly brown chicken feathers.

All my starlings left in the next day, and I never saw or heard them in the vicinity again. But I saw the pattern of parenting that takes place after they fledge in many others. The adults at first keep track of their fledged young and go to feed the one that screams the loudest when they are near. In a day or two the hungriest, the loudest screamers, come to the parent instead. Soon all of them scream and come for the parent that comes back with food. From there it is only a small step that one or two, and then the whole crowd of young, follow a parent. The young get agitated and sometimes even peck at the parent if it does not feed them (that is, if it does not find food quickly enough), and quite often the parent in turn makes an occasional jab at one of its young that is interfering with it as it races over the ground either to find food or to outrun the young or both. Eventually the young are so close to the parent that they sometimes find the food before the adult can pick it up to give it to them, and from then on all family members soon forage together. Gradually the families coalesce into ever-larger swarms that stay together through the rest of the summer, fall, and winter, and then dissolve in the spring.

Phoebes, unlike starlings, have open-cup nests, and they are one of the few species of passerines that regularly re-use their nests. By building on a small ledge in a cave, or house, the nest is not destroyed by the weather as most bird nests are. Furthermore, because the nest is often on a rare perch such as a nail on a wall or a shelflike projection, the old nest becomes a functionally useful platform for the next nest. Could that have implications for nest hygiene?

I had not paid much attention to the phoebes' nest hygiene, probably because almost invariably I never saw even a speck of feces near most nests—until I documented something strange. At the ten observed nests (from 2006–2008) the birds had two broods of three to five young per summer. In the first broods there were only zero to five fecal pellets left under nine nests, although one had accumulated forty-five pellets. This was as expected. However, on the day that the young of the second broods fledged there were thirty and up to 120 new fecal pellets (mean of 64.4 per nest) accumulated under the nests. It looked as though the birds planned ahead by cleaning up after themselves only after the *first* nest use, when they would use the nest again.

It required sixteen days from hatching to fledging of the second brood, but the accumulation of feces under the nest did not start until day ten, and it accelerated from then on. In one nest the young left on average 4.7 droplets per bird per day during the last three days. In another the counts of droplets per day per bird increased over the last three days from 2.6, to 4.8, to 7.0 on the last day. These results convinced me that as with the hole-nesting starlings, the parents at first removed the feces, and later they stopped and allowed them to accumulate under the nest. Most other passerine birds practice stringent nest hygiene at all times.

Flexible behavior such as the phoebe's demands an explanation. Is it because there are nutrients left in the feces early in the year that are absent later, so that nest hygiene is mainly a means of feeding themselves? Curiously, I have seen only three other local birds who abandon nest hygiene. They include ravens, whose young when they are large practically whitewash the nest and its surroundings. Barn swallows also allow the feces to accumulate under the nest. Both of these species tend, like phoebes, to nest at cliff or equivalent sites, so their lapses of hygienic behavior might reflect their relatively safe nest sites. In Hawaii on Mauna Kea, a honeycreeper leaves a huge ring of feces around its nest (Ellen Martinsen, pers. comm.). The island has traditionally been free of mammalian predators, and perhaps nest hygiene is not required for the safety of the young.

There is one other bird that is unusual in not cleaning up after the young—namely the American goldfinch, which nests in the open on branches. This is the latest-nesting passerine bird. It waits until seeds, such as the thistles, are available for itself and its young. The European goldfinch has a similar diet and life cycle and a study of this species shows that the adults eat their young (but not old) nestlings' feces as an energy source.[24] The young nestlings' feces have a high energy content, but as they grow older and their digestive efficiency improves, the energy content of their feces becomes equal to that of the parents', and at that point the adults cease "nest hygiene." The nutritional hypothesis for nest hygiene is also supported in American crows and Florida scrub jays, in whom the females, who are more nutritionally stressed than the males, eat more feces than the males.[25]

*Post-fledging care.* One of the potential problems bird parents face is to get their young to leave the nest. Often that leaving is seemingly delayed. For example, dusky shearwater young of New Zealand stay in their nest burrows for over two months, and by that time they weigh twice as much as their parents. The method to get them out of the nest is simple: the parents stop feeding them. After one week without food the young venture to the burrow mouth, and after starving for another week they finally take to wing. Raptors usually use the carrot rather than the stick, and instead of flying directly onto the nest to drop off their kills, they perch nearby and induce their young to come and get it. Peregrine falcons also present the food in the air, either by handing it off or dropping it. A great horned owl that I raised free in the Maine woods accompanied me as I went hunting squirrels for it. It appeared to know when I went hunting because it only then ventured from the "nest" area with me (not, for example, when I went to the mailbox). It watched me intently and instantly attacked a squirrel when it fell from a tree, but even two months past fledging it never tried to attack a healthy squirrel in a tree, nor would it go hunting alone. Similarly, many precocial birds such as turkeys, for example, stay close to their parents and observe what they pick up, thus learning what to eat.

In other birds, such as some corvids, the young may be dependent on their parents for a month or more, and during this time they also follow their parents and learn from them. Common ravens fledge in early May, then remain in the nest vicinity for several weeks as the parents continue to feed them. Gradually they meet the parents halfway when they are fed, and later they follow the parents around continually, learning what to feed on and where. By late July they often scream almost continually, and the parents act as though they are actively evading them, flying for miles and ascending high, soaring in the clouds, and also flying low through the trees. By late August the young have left their parents or vice versa and instead joined up with other young with whom they pool information on where to find food. In some of the large waterfowl such as geese and swans as well as cranes, the parents continue to serve their offspring not with food but with vital information. Family groups join up in large flocks in late fall, and the adults lead the way on the migration routes. Without their guidance the young would not know where to migrate and would not likely survive the winter.

All in all, parenting by whatever social arrangements can be a tough job. It is said that it is easy to become a parent, but difficult to be one. And many birds illustrate that truism well. Some have become supreme specialists of the first and totally incapable of the second, and others are up to both tasks. Those who shrug off the duties of parenthood are well known, though not always well understood. They are the cuckolders and the brood parasites.

# 11

CUCKOLDS,

CUCKOOS,

COWBIRDS, AND

COLOR CODES

*Nature does nothing merely for beauty; beauty follows as
the inevitable result.*

John Burroughs, 1877

*JUNE 22, 2009.* THE CHIPPING SPARROW
nest that I have been keeping track of for two weeks is almost
ready to fledge a cowbird and two of its own chicks. The baby
cowbird looks huge next to the two young sparrows in the cup
nest lined (as is typical) with fine animal hairs. While the sparrow had laid its four beautiful green-blue eggs into this nest,
an adult cowbird had inserted her own white egg, which is
blotched with brown speckles. She had then removed one of

the sparrow's. Would the sparrows have deserted their nest if the cowbird female had removed more of the sparrow's eggs? The sparrow could not have had an idea what even its own eggs might look like until it had laid some. It also would have no idea what its own babies would look like, and it accepted both. One of its hatchlings had a hugely larger mouth than the others. Its bill was edged with white, rather than with yellow as in the other three. In several days one of the smaller babies was missing. There had been cold rainy days; it probably starved because the parents would have fed the healthiest ones—those that presented the biggest mouths. The cowbird's was the one that got most of the food. And now it was big and the others fewer and small.

*May 23, 2007.* Tree swallows were flying in and out one of the bird boxes in front of my Maine cabin, but when I opened the box I saw that the nest foundation had been made of green moss, a material that tree swallows don't ever use. Chickadees do, however, and indeed the nest contained not only five pearly white tree swallow eggs but also two (brown-spotted) black-capped chickadee eggs. As is typical of tree swallow nests, the nest lining was composed mostly of light-colored feathers on top of the moss, so the swallows had apparently taken over a nest started by chickadees. The eggs of the two species were on the same nest lining, so the swallow and the chickadee had probably laid eggs alternately but on the same days. The swallow finished laying first because the chickadee would have been unable to enter the nest after she started incubating, and so the nest contents were then left to the swallows.

*May 27, 2008.* A small group of least flycatchers is nesting locally this year near a bend in our dirt road, and I go there periodically to check on them. At one nest I have been watching I noticed for the first time a bird sitting on the nest. Almost always incubating birds sit very still and don't seem to move a muscle, but this one was jerking its head here and there (though not at me, far below her). Suddenly it jumped from the nest and flew off into the woods. Strange, I thought—why was this bird so agitated and why leave the nest in such a hurry and fly so far? A second later I saw a vigorous chase between two least flycatchers, and then almost immediately one of them flew unhesitatingly back and onto the nest, settling deep into it and remaining perfectly still. Was the nervous bird a neighboring female who borrowed this nest to lay an egg and was then discovered and chased by the nest owner?

In early May at the cattail marsh and beaver bog near my house, the female red-winged blackbirds return and there is a frenzy of activity as they start to nest and lay their eggs. I see one crouch and flutter and a male jumps on and copulates. Some other females are near what seem to be other pairs, but I cannot be sure; the females move around a lot and seem to be ignored by the males. On the other hand, the same male will suddenly chase a female who makes every attempt to evade him; it looks like he is trying to chase her away. Is she from a neighboring territory? Male redwings copulate with more than one female if they can. Could the invader be attempting to lay an egg into the nest of one of his females? It makes sense for the females to try to dump eggs into neighboring nests, and it makes sense for a male to try to chase them away because a female from a neighboring territory is likely to be fertilized by a different male, so he loses genetically if his mates' nests contain eggs fertilized by other males. But I have no evidence that this occurs.

In the hurly-burly of this bog full of birds it is impossible to know who belongs to whom. In a simpler social context it might be easier to detect invaders, although home invasions might occur far less often. In a tiny boggy space bordered by woods and fields where I had always seen only one male and one female redwing, I searched and was confident that there was only one nest. However, one day while the female there had started to lay her eggs I saw the male take off and vigorously chase a second female away from his territory clear across a field. I immediately checked the nest and found that although it had held four eggs the day before, it now had only three. Had the male caught an intruding female in the act of removing an egg and chased her away before she could replace it with one of her own? In this case, however, the intruder (if it was one) was successfully evicted, because the nest stayed at three eggs even though the local redwings almost always lay four eggs per clutch.

It is estimated that over 90 percent of all birds pair up into couples during the breeding season and often beyond, and some pair even "for life." Birds (excepting all barnyard roosters, of course) have therefore traditionally been seen as paragons of monogamy because they are so often together as couples, especially during the breeding season. Ironically, some of the babies in their nests are often not their own, or at least they are not genetically all theirs. Many are fathered by a male other than that of the pair. For example, in one study fully half of the nests of the least flycatcher contained young whose fathers were from a neighboring pair.[1] Some of the young in a clutch may also be genetically unrelated to both parents and may even be of a different species. The stories behind these observations are varied and complex, and have generated a large body of literature on the fascinating topic of bird mating and nesting behavior, some of which are crass and unsavory if referenced to human ideals.

*Cuckolds by extra-pair fertilization.* Birds don't have a penis, although some have a semblance of one in one, or sometimes two, extensions of the cloaca. These birds include ducks (who have a corkscrew-like organ that can spring instantly into action) and ostriches (in whom the extension is red and about thirty centimeters long). In most birds matings consist of a few seconds of cloacal contact that therefore requires close cooperation between the pair. During mating, the female holds her tail aside so the male can bend over backward while balancing himself on her back. Matings occur in a very narrow time window, just before the eggs are laid and when the female is fertile. Somehow the male is apparently apprised of her physiological readiness.

Those matings with neighbors outside the pair bond, either extra-pair copulations (EPCs) or extra-pair fertilizations (EPFs), or sometimes just extra-pair paternity (EPP), are extremely common and adaptive for both partners despite conflicts of interest.[2] Extra-pair copulations and fertilizations are not equivalent. For example, in Humboldt penguins 19 percent of males and 31 percent of females engage in EPCs but DNA fingerprinting shows that none of these result in fertilizations.[3] Nevertheless, typically a third of the young of migrant passerines in North America have a biological father who is not their social father at the nest, and such a high prevalence of EPFs makes it unlikely to be a product of just random chance. The consensus is that in males the phenomenon is in many species an evolved strategy of becoming a parent without parenting. In most species, however, the extra-pair copulations benefit females more than males, and females often solicit copulations, although not necessarily always for fertilization.

In birds, one copulation can go a long way. One copulation is all that is needed to fertilize a whole clutch of fifteen or more duck eggs. And the sperm can last a long time. In hornbills, the female is entombed inside a tree hole a week before she starts to lay eggs. Nevertheless, the sperm from the copulation before she seals herself in fertilizes her eggs two to three weeks later. EPCs by males can, however, be costly despite their potential, because socially polygynous males who leave their territory for EPCs are themselves frequently cuckolded.[4] That is, his cuckolding gain at one end can be lost at the other because cuckolding detracts from his mate guarding.

For females, who are limited to laying a certain number of eggs, the potentially mean advantages of EPCs are thought to be "good genes" or heterozygosity of offspring or different genes that are more compatible with her genotype than the mate's to increase the viability of her offspring.[5] In the Australian fairy wrens, for example, where the young tend to stay in the parent's territory and pairs are often related, EPFs are rampant and may involve over 40 percent of the offspring; both males and females may copulate with a half-dozen others outside the pair.[6] The EPCs by these wrens may not only increase reproductive output by the males but also reduce inbreeding depression. In some species of passerine birds it is the males and in others the females which make most of the off-territory forays. These result in copulations that may be more likely to fertilize their eggs than their social mate's. Female collared flycatchers who have a small white forehead patch may sneak a tryst with a neighboring male if he has a larger white patch and thus make the sperm of the latter be more likely to fertilize them.[7] Black-capped chickadee females are also highly selective in their EPC partners and prefer high-ranking males outside their territory, even though pair bonds with their regular partners may persist for years.[8] Although the advantage of EPP sought by females is probably largely "good genes," they may gain direct material benefit as well. For example, as determined by Elizabeth M. Gray from the University of Washington, female red-winged blackbirds apparently trade food for sex. Results of feeder experiments showed that a male allowed females to feed on his territory if they mated with him, but chased away those who did not. These females then also had greater fledgling success. That is, for these females extra-pair copulations paid off materially as well as genetically.[9] Similarly, spotted sandpiper females may use EPCs to acquire another male to care for their eggs while starting a clutch with a different male.

A potential cost to a female seeking EPCs is that she risks losing some of the male's paternal investment in her own young because he presumably would only help his own. In one test involving Eastern bluebirds and tree swallows, Bart Kempenaers and co-workers tried to reduce the mate's certainty of paternity by temporarily re-moving the fertile females during the egg-laying period to see if their males would then invest less effort in the young. No such effect was seen; the males still helped her as before in feeding the young. In-stead, in both bluebirds and swallows the males immediately copu-lated when the pair were reunited. The authors argue that perhaps the females restored the confidence of their males by their soliciting and then accepting these copulations.[10] They were not extra-pair cop-ulations but rather extra-within-pair copulations—not needed for fer-tilization though potentially increasing the chances of restoring pa-ternity of subsequent fertilizations through sperm competition. In general, males accompany their females during their fertile period and guard them jealously from EPCs. Such mate guarding may look like an intense bonding based on "love," and the behavioral responses are equivalent to those that might be predicted from the emotion la-beled as "jealousy" in humans. In a colonial sea bird, the blue-footed booby, the males as in most other species of birds closely accompany their mates during their fertile period before they lay their eggs. How-ever, when the females' mates were experimentally removed during this time and then returned just before egg laying, 43 percent of the males expelled the first egg his mate laid from the nest. They did not expel any eggs if they had been removed from her just prior to her fertile period.[11]

Wood thrushes are solitary forest birds who maintain territo-ries. However, during the females' fertile period the pairs engage in an extraordinary number of off-territory forays.[12] After the egg lay-ing, when the females are infertile, they no longer wander and are not accompanied by their mates. The authors of this study therefore con-cluded that the females were likely either searching for opportunities to dump eggs into other nests, or seeking EPFs, and the behavior of the males was therefore mate guarding. To distinguish between the two possibilities, they conducted a genetic analysis of 151 nestlings and found that all of them were related to their social mother. Egg dumping by the birds in this study was therefore highly unlikely. Their off-territory forays, however, resulted in an equally unprecedented low incidence (6 percent) of EPFs, suggesting that the male's mate guarding is probably a very effective counter-strategy to her EPF attempts.

One of the perhaps more bizarre attempts by the male to counter a female's EPFs is found in a European sparrow, the dunnock. When a male in this species is cuckolded, his mate may present her cloaca to him, expelling a droplet of sperm that he then picks out and eats before replacing it with his own.[13] Male dragonflies, who are ahead of birds in the mating game by hundreds of millions of years of evolution, have these matters honed to exquisite perfection. They have "sperm scoops" directly on their penises that automatically remove the deposits of previous inseminations before they deposit their own. Then, as insurance or counterstrategy to sperm scoops, the males of many insect species add a sperm plug, or alternately become one by staying attached to their female with special claspers apparently designed for that task. They thus stay monogamously mated for hours or days and possibly also "for life" (since they don't live long). A method to prevent fertilization of a male's mate (one to my knowledge tried only by ornithologists in a lab study) permitted the female to be fertilized by another male: In a study of red-winged blackbirds, it was found that some of the females laid fertilized eggs despite their mates' vasectomies.[14]

Unlike insects and unlike most mammals, birds have nests, so there is an additional mechanism of cuckolding: instead of males dumping sperm into females' cloacas, females can dump eggs into the nests of other pairs. The two mechanisms are not necessarily easy to distinguish except by direct observation, or by genetic analysis. A study of a population of white-crowned sparrows in California distinguished between EPFs and egg dumping and provides an illuminating discussion of the EPF strategy. Paul W. Sherman and Martin L. Morton determined that at a minimum a quarter of their sparrow nests contained young that could not have been conceived between the attending pair.[15] Since no mismatches in genotype (using electrophoretic data in this study) between the females and the young were found, therefore all of the eggs in the nest had been the females' own, and the genetic mismatches in this species were therefore due to male cuckolding by EPFs and not to egg dumping. Many nests contained more than only one chick fathered by a male other than the social father who was defending the territory and feeding the young. Furthermore, the total number of chicks with mismatched genotype (14 percent) under-represents the actual total extent of EPCs because matings with the same genotypes of the putative parents would not have been detected.

A previous study showed that colonial bee-eaters also have a low frequency of foreign young, and Sherman and Morton were therefore surprised at the high incidence of EPCs in their sparrows, although it was nearly identical to that of indigo buntings.[16] Sherman and Morton therefore suggest that since both the male sparrows and buntings expend considerable effort in singing and in vigilance to attract other females for EPFs and to repel rivals, they cannot simultaneously effectively guard their own females from cuckoldry. The male bee-eaters, in contrast, engage in intense mate guarding during their females' fertile period. Furthermore, the males then can and do copulate with them frequently, thus increasing the chances of their paternity not only behaviorally but physiologically by sperm preponderance in those cases where there might be EPCs.[17]

*Cuckolds by egg dumping.* The females of many birds, because they lay several eggs together in nests, have the opportunity to accidentally (or deliberately) use another's preformed nest. Mixed species clutches are thus possible. In such ducks as the lesser and greater scaup, mixed clutches have resulted in which one lays in the nest of the other, but the other female is not evicted. Instead the two incubated side by side and in contact with each other.[18] Sometimes the respective females incubate the shared clutch in one nest alternately, as in the case of two northern cardinal eggs in the nest of an American robin.[19] Usually, however, when two females lay in the same nest, one of them leaves or is evicted by the other. Such behavior becomes the basis for an evolved strategy of deceiving other females (and their mates) by depositing their eggs into their nests and hence relinquishing their care. Anyone who has ever raised chickens is familiar with egg dumping. Chickens preferentially lay their eggs into a preexisting nest that already has another's eggs in it, so to induce chickens to lay into a particular nest, one puts an egg into it. Egg dumping is common in others of the chicken/grouse/quail family (Phasianidae) and also in waterfowl, coots and their kin, and colonial birds, such as many weaver birds and swallows. Undoubtedly it exists in many others. Although egg dumping is at times incidental or accidental, such "mistakes" have provided the raw material to shape the evolution of ever more refined and complex arms races between nest parasites and their hosts.

Some of the most obvious egg dumpings represent mistakes because they have costs and no conceivable benefit. One extreme case in which one species "accidentally" dumped eggs into the nest of another species was described and shown to me in a photograph by the ornithologist Carlos D. Santos. In the Cosumes River Reserve near Sacramento, California, Santos had found a typical ground nest of a spotted towhee containing four typically blue towhee eggs with small reddish-brown spots, but in among this full towhee clutch were also four conspicuously larger California quail eggs, which are white and marked with large dark brown blotches. Three days later the nest contained the same number and mix of eggs and it was being incubated by the towhee. Although the egg number of the towhee was typical for its clutch, a complete California quail clutch normally contains at least twelve eggs, with one report of up to twenty-eight (likely the result of more than one quail using the same nest). Most likely the quail laid eggs into the towhee nest during the days when the towhee was also in the process of laying and thus away from the nest for most of the day. The quail was excluded when the towhee finished laying its clutch and started to sit on the eggs. Such egg dumping is easily detected by us because of the huge mismatch in the size and colors of the eggs. But if it occurs within the same species, where it is presumably much more common, one could not "see" it with the naked eye. In the above case the quail made an unfortunate mistake and compromised her reproductive effort, but had she dumped those eggs into another quail nest she would probably have benefited from her behavior.

Egg dumping within the same species is difficult to document because there are no or only slight color differences between the host bird and parasitic eggs. Egg dumping is still obvious, however, when the numbers of eggs per nest are conspicuously larger than the known clutch size. In one study, several female wood ducks laid their clutches into the same nest box, and there were often far too many eggs in some boxes to be incubated, so only a few young could hatch. Extreme egg dumping was the result of a shortage of nest sites.[20] In another case in Michigan, European starlings had usurped all available nest boxes so that northern flickers had none. A bird box was put up that was filled completely with woodchips so that the starlings could not use it but flickers could excavate and thus claim it for themselves. This box ended up with sixteen flicker eggs that must have been deposited by several females.[21] However, if egg dumping is costly to the recipients, then counter-strategies would be expected to evolve to limit the behavior by the hosts.

In the expected arms race between the egg dumpers and their hosts, the dumpers should improve their abilities to insert their eggs while the hosts should improve their abilities to prevent the parasite eggs from being inserted. There may be winners and losers. But it is not always obvious which is which. "Winning," "losing," and "parasite" are words with anthropomorphic implications that are not necessarily always accurate. They imply conflict, which is not as clear as the terms imply. For example, in the wood ducks, who often lay in one another's nests when suitable tree cavities are rare, there is usually only one female who ends up incubating. Is she the "winner" or the "loser" in the interaction? She could be considered the "winner" of the nest site if they fought for it and she remained, but she would thereby become a loser in caring for the others' supposedly parasite eggs, especially if she accepted many and therefore lost her own.

In my previously mentioned example of finding chickadee eggs in my bird box with a tree swallow nest I expected both species' eggs to hatch and the young to grow on a common diet of insects. I expected to see a mixed-species clutch emerge. However, when I returned two weeks later, I found five half-grown young swallows in the nest box but no young chickadees. One rotten chickadee egg was at the nest periphery. It had been rolled out of the nest mold, yet it contained a nearly full-term embryo so it had been incubated for awhile. I did not at first see the second chickadee egg, but since there was now no danger of the birds abandoning the nest, I temporarily removed the young swallows and made a thorough nest inspection. I found *two* more chickadee eggs underneath the feather nest lining. Since I had previously seen (and photographed) only two chickadee eggs, both on top of the nest lining, I conclude that the swallows had discriminated against all of the three available chickadee eggs by shoving them out of the way.

The egg dumping had not gone well for the chickadees. However, sometimes inter-specific egg dumpers are more lucky, cleverer, or sophisticated through evolutionary programming. An example of some rails suggested the latter. I had stopped my car along the marsh at the end of Shelburne Pond near our home in Vermont, as I commonly do, and heard both sora and Virginia rails calling. It was nesting time, so I waded into the marsh and indeed found a rail nest. It held only two eggs. I checked the nest again five days later, and it then contained ten eggs that were being incubated by a sora. Furthermore, and to me even more puzzling, an empty rail egg shell lay in the mud several meters distant. Eight or nine eggs could not have been laid by one bird in five days!

All eleven eggs were similarly cream-colored and spotted with reddish-brown and gray-brown spots and blotches, so I at first assumed they were all sora eggs. Not until I looked at the eggs more closely did I notice that although most of the eggs of the clutch were of uniform size, two of the ten were larger than the rest; at least two females had used the same nest, and one of them may have been a Virginia rail since that species has slightly larger eggs than the sora's. The broken egg, along with the vocal displays I had heard near the nest area, were clues that there may have been conflict and that this egg dumping was not "accidental." Apparently there had been egg dumping but also some discrimination. Indeed, this egg dumping, if it can be pulled off, is likely adaptive, because the young of the two species of very similar diet and natural histories would have been easily served by the same parents.

Egg dumping in other marsh-nesting rails (common moorhens and coots) has been documented, where it occurs within the same species. These rails are semi-colonial, and despite the similarity of egg markings between different individuals, coots reject many conspecific parasite eggs by pushing them to the nest periphery.[22]

In a study of barn swallows in Denmark, Anders Møller also found sometimes more than one new egg appearing in a swallow nest per day, although individuals lay no more than one egg per day.[23] Therefore, two different females had used the same nest. He noted that although barn swallow eggs vary greatly in spotting pattern between different females, when two eggs appeared in a nest in one day, then one of them was usually slightly different. Intra-specific egg parasitism occurred in 17 percent of nests, and it was more frequent in large than in small colonies. Parasite eggs were usually evicted from the potential hosts' nests if they were deposited before the host had laid eggs, but after the first egg was laid by a host swallow then parasite eggs were no longer rejected. It appears, therefore, that the swallows either cannot tell their individual eggs apart, or they can tell them apart but invoke a rule of not ejecting anything from their nest once egg laying starts, on the evolutionary "logic" that it could be one of theirs. However, they can be certain it is *not* their own if it appears before their own are laid.

Female barn swallows visited their neighboring nests during the owners' fertile period, perhaps to ascertain egg-dumping opportunities.[24] Apparently the defense for it is nest guarding, since those females who were most parasitized were the ones that spent the least time at the nest. Nest guarding is not just free time—it takes time away from foraging, and since all must forage there is always an opportunity to slip an egg into a neighbor's nest.

Birds that randomly dump eggs into the nests of other birds are likely engaging in a costly behavior. First, their eggs will probably contrast with those of the presumptive host and could be ejected or discriminated against. Second, even if accepted and incubated, then the match-up between the parenting that this host provides and what the hatchling would need may be poor; a towhee cannot care for quail chicks. On the other hand, dumping eggs into a nest of its *own* or a similar species is potentially ideal; egg matching could be perfect and nestling requirements would be close to identical. Therefore, such egg-dumping behavior may be common but if so it is difficult to detect. For example, canvasbacks, redheads, and ruddy ducks, who all have cream-colored unmarked eggs and inhabit and nest in open nests in marshes, often lay in each others' nests. Social situations, such as in colonial swallows, also allow for parasitization within the same species. In cliff swallow colonies about a quarter of the nests have brood parasites.[25]

Since a bird should be unable to identify its own eggs before laying and then seeing them, there is a potential conflict between learning to evict nonmatching eggs and learning (imprinting) on just-laid eggs. For example, if one were to prevent learning by a catbird of its own bright blue-green eggs by replacing each as it is laid with a speckled cowbird's egg, then after she has accumulated a "clutch" of cowbird eggs in her nest, she would predictably reject one of her own eggs rather than doing the reverse, as happens in the wild. Selection for evicting differently colored eggs could be costly in those birds that lay variably colored eggs. Indeed, common grackles and red-winged blackbirds lay variably marked eggs even within a clutch, and these birds do not reject cowbird eggs, whereas other grackle species that lay uniformly colored eggs within a clutch reject all cowbird eggs.[26]

Given the potential ease of parasitizing other nests of the birds' own species, there is therefore likely strong selective pressure on birds both to parasitize and to prevent being parasitized. The countermeasures could predictably involve color coding to make the eggs of potential nest usurpers stand out. For example, if intra-specific parasitism is rampant because a stranger's cannot be recognized and then differentiated from one's own, then a female who lays eggs of a "different" color than her neighbors would automatically have leverage to be able to identify a stranger's egg among her own. Her egg color mutation should have an advantage over the species' normal or common egg color pattern. That is, one would expect variety of color and coloring complexity to evolve within the species because the only way to have a neighbor's eggs be conspicuous is to have a different color pattern than theirs. Variety, however, is a function of frequency, and thus it all depends on what the others' colors are. Suppose your neighbors (regardless of species) try to pass off one their green eggs to you, adding to the existing population of green ones in your nest. You could avoid being cuckolded by having red ones instead. If you had the red mutation, it should spread because it would resist parasitism. On the other hand, if you have the common green phenotype, you can more easily parasitize because you have more potential hosts who would find it difficult to discriminate against you. Eventually an equilibrium would be reached in which both color morphs exist (or in which still other color morphs evolve). The benefits of parasitization equal the costs of being parasitized. Thus the variation of egg coloration between different nests may allow a bird to recognize a neighbor's egg in its nest, but it would also allow a neighbor to detect its egg in their nest. In Africa, a number of colonial weaver bird females lay individually specific color morphs all their lives.[27]

*Weaver birds and African safari.* It is a snowy evening in late December in Vermont, and I've pulled out an intimidating book that is redolent with powerful memories. The gray nondescript cover is stained and worn after having seen heavy use in the African bush for two years. On it is written in my father's familiar handwriting: "Pass."(Passerines), volume 2, the Passeriformes of *Birds of Eastern and Northern Africa*, by C. W. Mackworth-Praed, M.A., F.Z.S, M.B.O.U. and Captain C. H. B. Grant, Hon. Associate, Zoological Dept. British Museum, C.F.A.O.U., M.B.O.U.

This was the 1,177-page volume that my father must have received virtually fresh off the press in 1961 when we boarded a freighter, the *African Moon*, in New York City. About a month later, we arrived in Dar-es-Salaam, Tanganyika (now Tanzania). Not a day went by on our journey that we did not have it out to study the bird fauna we soon expected to encounter as we collected for the Peabody Museum at Yale University. I was 21 years old and an eager bird naturalist, and by instinct and practice a hunter from age 8 or 9. After we landed, the book then accompanied us during our year in Tanganyika's Pugu Hills, the Usambara, Uluguru, and Ukuguru Mountains, and also Mount Meru and Lake Manyara and environs. The book was our bible and the authors were, according to their hefty title, obviously qualified gods. But the book content that interested me now was inconsequential to me then.

Forty-eight years have passed, and my memories start flooding back in a torrent. But I'm not now opening this book for memories. I've retrieved it from our attic for something to which I scarcely gave a thought over the entire ocean voyage and the year beyond: I'm wondering about the color of the eggs of weaver birds and how it relates to their natural history in terms of adaptation. All I had previously known was that in some species of colonial weaver birds (such as the weavers *Ploceus taeniopterus* and *P. cucullatus*), each female lays a clutch of eggs that is distinctively different from the eggs of most of the other females in the same colony.[28] In this case, since the two to three eggs of a clutch are nearly identical, yet different from others' who have a half-dozen different color morphs, a bird that has a distinct egg color morph provides a unique background or contrast to its own eggs. Mismatches would identify an egg from another female of the same colony that might slip into the wrong nest by mistake or by deliberately trying to dump an extra egg. In contrast, the eggs of other African colonial weaver birds (such as Speke's weaver and the sociable weaver) are an intense robin-blue and show almost no variability. These facts suggest the dynamic nature of color patterns inscribed on the birds' eggshells, reflecting the strange game of stealing parenting duties that birds have long engaged in.

Previously I had used my big gray book mainly for its identification keys and range maps and perhaps habits. Information on nest and eggs then seemed trivial, but now this information—and the book was full of it—had suddenly become intensely interesting. I was astounded how much information on nesting and eggs the two authors had amassed. It was information that must have represented the patient observations and recordings of hundreds of birdwatchers who are now anonymous and the fruits of their interests almost forgotten. It probably was the product of the hobby of egg collecting, which through abusive trade of rare species has been discredited and largely forgotten.

I looked at the weaver birds first. They, in their often noisy and conspicuous colonies, are one of the most prominent birds found in Africa. At random, I came to the yellow-backed weaver. The nests are described as "roughly kidney-shaped and without a funnel." They are built "mainly by the male, the female helping with the nest lining." The nests are built of coarse grass with an inner lining of fine grass, and they are suspended from trees, bushes, elephant grass, or papyrus. The eggs, usually two, are "most variable, white, pink, brown, chocolate, liver colour, terra cotta, or various shades of green, with or without brown or purplish spotting, and rather hard shelled." The birds are "very gregarious and nesting in colonies with other species of weavers, usually over or near water. They often start a colony and desert it suddenly after a few days."

Undoubtedly a trove of biology lies hidden here. There is apparently as much egg coloration in this one species as in almost all the hundreds of passerine species of North America! And the bit about "rather hard shelled" was intriguing as well—it immediately suggested an adaptation to protect against nest disturbance and hasty egg dumping by nest parasites and possibly by colony mates? Would all weavers have such variably colored eggs? Not at all. Turning back a page, I came to the little weaver, a bird that on the color plate looked similar to the yellow-backed weaver. But the eggs of this species are "white." I was stunned. What could be different about these two species that might explain their contrasting egg coloration?

Both are "common and gregarious." Only one major difference seemed obvious from the descriptions given; the latter, the little weaver, "has a strong tendency to nest near wasp nests," and "the nests are shaped like a spherical pipe with a stem pointing downwards, the entrance funnel running up towards the top of the bowl, and there is a sort of gate inside to prevent the eggs falling down the tube." I was now excited because I had a hypothesis: maybe this weaver accomplishes with its nest what the yellow-backed does by egg color. I stopped to make a watercolor sketch of both weavers and their nests and eggs to fix both firmly in my mind, meanwhile thinking about the contrasts. My first thought was that the little weaver's nest is a defensive structure designed to limit access; the long entrance funnel—the pipe stem—would be a formidable barrier. This bird is quite small and the entrance would exclude most others. In a sense, the nest is a reverse woodpecker nest structure—one woven, the other excavated. Woodpecker eggs are always white. Similarly, the brown-capped weaver also has pale, unmarked eggs and also has a long entrance funnel into its nest. There is no need for color coding to differentiate the birds' own from strangers' eggs. The "sort of gate" at the top of the entrance funnel would not only keep eggs from falling out but also keep parasites out. The wasps would presumably provide defense as well, one that operates even before the door. In contrast, the yellow-backed weavers' nests are practically open. Since this species nests in mixed-species colonies, the "most variable" colored eggs should make it easy for each female to detect nest parasites not only of her own but also other species.

I would like to be able to report that this explains all. It doesn't. I went through the descriptions of all the weavers in Praed & Grant, and it was indeed true that most of the weavers nesting in dense colonies had very variably colored eggs (*Ploceus cucullatus, P. vitellinus, P. velatus, P. capitalis, P. castanops, P. xanthopterus, Hyphanturgus migricollis,* and *Xanthophilus xanthops*). However, a "usually solitary"-nesting weaver *(H. nigricollis)* with a long nest entrance funnel also has "very variable" eggs. Unlike the others, this species lives in thick forests in low-lying country, suggesting possible additional or other variables. For example, the nest of this species has "a small entrance hole at the side, but the presence or absence of a porch varies with the locality. In Natal there is usually a porch, in Nyassaland no porch." Similarly, the buffalo weaver is reported to have "pale blue eggs" in Darfur, "blue" eggs in Uganda, and "white or pale green blotched or streaked in gray and brown eggs in the rest of Africa" where it occurs. Undoubtedly evolutionary precedent also plays a major role, so that selective pressures keep changing and some features change faster than others, as well as varying from place to place.

*Egg recognition.* Most eggs of the corvid family are bluish-green, blotched and spotted with gray, purplish, brown, and black. From a distance they blend in with their environment of vegetation in their nests and the foliage of trees. Often they are variable among clutches and even within the same clutch. Do these colors mean anything to the female that laid the eggs? I once had an opportunity to find out at a raven's nest on a small cliff in Vermont. It was an unusual nest in that I could easily reach it with a small ladder, and whenever I came near it the raven left and flew up and over the hill. This seemed like an excellent experimental opportunity to find out if these birds, who are reputed to be intelligent, recognize strange eggs among their own. Working quickly, I could put an egg into the nest and be confident that I was not being watched. I meticulously colored a small chicken egg (near the size of the raven's) to mimic a raven egg and placed it into the nest. To my surprise the fake raven egg was accepted. So I exchanged it for an unpainted chicken egg. The result was the same. Ravens are alert and keen-sighted as well as intelligent, so I felt they would surely notice a hen's egg that I painted crimson. Yet I found later that she accepted (incubated) not only crimson hens' eggs but even flashlight batteries, potatoes, film canisters filled with sand, and rounded rocks that I put in (and removed) sequentially on successive days.

In retrospect, I didn't know if the raven could identify her own eggs, but acceptance did not mean lack of recognition. I had an anthropomorphic assumption: I assumed they recognized the difference between a hen's egg and one of their own because we can easily identify and distinguish such objects, but what may be simple to us need not be easy for them. However, why should they throw a strange "egg" out, given that their nests are probably never parasitized; they would see a stranger coming from far away? Nevertheless, the female's alarm calls and sometimes hesitancy to resume sitting on the nest after I visited suggested that she indeed recognized them. In retrospect, the likely explanation for the ravens' behavior is that they have no reason to pay much attention to what their eggs looks like. Even if they did recognize the objects as strange, any egglike object they would reject would normally be their own, so the penalty for making a mistake is great while that of accepting even a very strange-looking "egg" is vanishingly small since normally no other bird's eggs, or potatoes, end up in a raven's nest. Ravens sometimes have clutches of up to seven eggs that they incubate successfully, although almost always there are only four or five, so there is almost always room for one more egg, though not necessarily one more young.

Only after the eggs hatched did the raven parents start nest hygiene behavior—they picked up and threw out empty eggshells, regurgitated pellets, fecal droplets, rotten eggs, and all the fake eggs that I put in. I suspect now that the raven's tolerance of foreign "eggs" is probably not based on either inability to discriminate or lack of intelligence. It is due to its natural history and instincts to resist damaging its own nest contents. A stark contrast that illustrates this point is provided by common murres and their close relatives.

Murres are colonial seabirds that build no nest and lay a single egg per clutch onto a bare ledge close to the eggs of other murres. Murre eggs of any one colony are highly diverse. In common and thick-billed murres and razorbills the eggs of different individuals of different species have been described as varying from white, cream, buff, reddish, blue, greenish, brown, and even black.[29] Eggs may be unmarked or with very few markings, or they can be covered with scribblings, spots, banding, and blotches of brown and black. The extinct great auk also laid only one egg per clutch and nested, similarly to murres and razorbills, in dense crowds. It had white eggs with bold markings of brown and black that were also said to endlessly vary in pattern. Each murre's unique egg color pattern is learned by the egg's owner, and ability to discriminate is used by them to claim their own egg in a crowd of others.[30] The birds know which ledge their egg is on, but switching the locations of different eggs on that ledge does not prevent the birds from finding and then incubating their own egg. Auks and razorbills have only one egg per clutch, so as long as they identify it they can be assured they are not wasting their parenting efforts raising another's young.

*Cuckoos.* Different cuckoos have one thing in common: their inner and outer toes are directed backward and the other two face forward. Otherwise they are quite variable. In North America they include the monogamous roadrunner, communal Anis, and the yellow- and black-billed cuckoos. The black-billed cuckoo forms pairs that build a nest and care for their own young, although in times of great food abundance when they can lay more eggs, they occasionally unload their greenish-blue eggs into each other's nests as well as into the nests of at least ten other species that have eggs ranging from pure white (mourning dove) to dark blue-green (catbird) and spotted (most of the others).[31] Anis build communal nests and several individuals lay their eggs into the same nest, so without additional information, it would be difficult to distinguish cooperation from parasitism.

The cuckoos' variation in social behavior and parenting care "may be unmatched among bird families of the world" as Robert B. Payne writes in his richly illustrated opus on this group of 141 species with a worldwide distribution.[32] No birds have a greater claim to fame as brood parasites than cuckoos, especially the common (Eurasian and North African) cuckoo. Some of the common cuckoos' fascinating natural history was already folklore when it was described by Aristotle, and it has been studied in great depth and reported in the scientific and popular literature, even up to the present day.[33] Most famous is the outcome of its co-evolutionary arms race with its hosts. Hosts have evolved to discriminate the cuckoos' eggs from their own, and cuckoos have evolved to camouflage their eggs by closely mimicking the color markings of their hosts' eggs. In one recent and illustrative study, for example, the cuckoos' "arms race" with one host, the marsh warbler, was investigated in Bulgaria.[34] The marsh warblers' fine-tuned ability to discriminate cuckoo eggs had evolved to the point where to the human eye the speckled cuckoo eggs look identical to the warblers' eggs. Nevertheless, the warblers reject about half of all naturally laid cuckoo eggs and nearly as many eggs (38 percent) of other marsh warblers experimentally put into their nests.

The eggs of the common cuckoo are in some cases (but not all) very closely color-matched to those of a specific host. The hosts, in turn, have evolved to become highly sensitive to the appearance of their own eggs and then to toss out from their nests any eggs that don't match those of their own clutch. Obviously there is a balance in which many of the (often closely matched) parasite eggs *are* accepted, or else the cuckoo would be extinct. It is now a matter of life or death for the egg a cuckoo lays to be properly color-matched to its host's eggs, as it is also for their hosts to detect them. Hatchling cuckoos will kill all the step-siblings in the nest. Strong selective pressure for color-matching by the cuckoo in the face of ever more critical egg discrimination by the host has reinforced the color-matching and hence forced strict host specificity in the cuckoos; cuckoos' eggs must match the eggs of a *specific* host species, yet this cuckoo *species* parasitizes many hosts. How did this system evolve?

In the distant past such color-matching would have been less important than it is now, since hosts were not yet selected to discriminate. A cuckoo could then lay its eggs in the nests of a number of different host species. Eventually some hosts evolved defenses, and thus they became poorer targets than others. But suitable host distributions varied, and so different color morphs of the cuckoos' eggs had different survival value in different areas. The behavior of using that host by individuals of perhaps a mutation could have spread quickly, especially if these birds lived in geographic isolation from other cuckoos. Cuckoos must specialize in parasitizing *specific* host species because egg coloration differs greatly among different bird species who have become "alerted" to parasitization and are now acute discriminators and active ejecters of unmatched eggs. At this point in evolution the common cuckoo species has formed genetically separated sub-groups called "gentes," which each parasitize a different host or hosts. At least sixteen different gentes have evolved in the same species, and each parasitizes one or a few hosts. Each gente has specialized with their own egg colors to match those of their own preferred hosts.

In the co-evolutionary arms race between cuckoos and their hosts one would predict that those potential hosts that have eggs that are uniform within their same clutch would succeed best in detecting a cuckoo's egg, but it is also expected that there is a premium for a species to have eggs that *diverge* from the species norm. That is, clutches colored differently from the normal clutches but similarly colored within the clutch should stand a greater chance of avoiding parasitization because the parasite's eggs would then be most easily detected.

Through time, it is expected that the distribution of gentes will overlap geographically and mix. Can they then still maintain their specific egg colors? It is possible only if there is reproductive isolation because a cuckoo's egg-color morph, like that of its hosts, is a genetically transmitted trait. The solution is imprinting. Different individual cuckoos imprint as nestlings on the hosts that raised them so that they then seek out that same species to parasitize when they reach sexual maturity.

Similar host specialization to that of the common cuckoo, where lineages of the cuckoo parasitize different hosts and mimic their eggs, is found in the Diedrick cuckoo of Africa. It parasitizes bishopbirds (who have blue eggs) and various species of weaver birds with speckled eggs of various background colors and a variety of markings. The hosts' counter-strategy to cuckoo parasitism is likely multi-faceted. Aside from identifying and rejecting cuckoo eggs, they would likely also have evolved individually specific egg colors which would greatly reduce the cuckoo's ability to color-match them.[35] Additionally, David Attenborough describes Diedrick cuckoos trying to enter through weaver birds' nest entrance tubes and becoming wedged in the nest entrance, thus failing to get in.[36] This observation suggests that the hosts may have responded with nest structure as a defense as well: their elongated nest entrance may serve as a cuckoo baffle since it is made just large enough for themselves, and cuckoos are generally larger than their hosts.

Selection against accepting a common cuckoo egg should be very strong since successful insertion and acceptance of a cuckoo's egg always results in the death of the host's whole clutch. Thus, the cost of a host keeping a cuckoo's egg is great. Therefore we would expect a host subject to cuckoo egg parasitization to eject every foreign egg it can detect, possibly even at the cost of incorrectly identifying an egg as foreign. Since a cowbird's parasitism only results in partial loss of a brood, and there is also a chance of losing the whole brood if the parasite egg-layer retaliates when its egg is chucked out, it is not clear whether or not the host should eject what might be a cowbird egg.

The great spotted cuckoo, who overlaps in distribution with

the common cuckoo in southern Europe, does not kill all its hosts' eggs or chicks and may represent an earlier evolutionary stage in the co-evolutionary progression of parasitization.[37] In Spain it parasitizes magpies, and its eggs closely match the color of this species. Further south, in Israel, it parasitizes pied crows. Still further south, in Africa, the cuckoo commonly parasitizes starlings, many of whom have domed magpie-like nests or breed in cavities. Possibly due to recent range expansion it now parasitizes other species, and egg matching is poorer since some starlings have unspotted blue eggs, others have light speckling, and still others heavy brown speckling.

As other nest parasites do, the great spotted cuckoo lays its eggs into the hosts' nests while they are laying and before incubation starts, but its incubation is shorter than its hosts—especially in large hosts such as crows—so the young have a head start, growing faster and fledging sooner. Unlike cowbirds and the common cuckoo, the great spotted cuckoo does not throw eggs out of the host's nest, nor do the host's young evict the nestlings. However, the cuckoos do damage host eggs, which kills some embryos. Furthermore, since the cuckoo gets the lion's share of the food, the growth rate of the remaining young is compromised.[38]

The cuckoo's eggs have thicker shells than the host's, so when the female cuckoo drops her egg into a magpie nest, it is more likely to break the host's egg than the cuckoo's. Manuel Soler and Juan G. Martinez of the University of Grenada in Spain tried to find out if the egg damage is the result of selection, and hence an adaptation, or if it is simply an accidental result of rapid egg laying.[39] They simulated cuckoo egg-laying behavior by dropping eggs into nests, and they found that this resulted in less damage to host eggs than in natural parasitism, indicating that host egg damage can increase the cuckoos' breeding success. Therefore, the cuckoos do something to exaggerate the potentially natural or accidental egg damage to their host's eggs.

Furthermore, in the experimental treatments the number of damaged magpie eggs did not depend on the number of magpie eggs in the nest, but when the cuckoos laid their eggs the number of damaged magpie eggs increased in proportion to the number in the clutch. These results imply that the egg damage is a result of adaptation and not just an incidental byproduct of rapid egg laying (although rapid egg laying could also be an adaptive response). Taken together, it appears that the sophistication brood parasites employ to bring young to adulthood without actually parenting themselves could be as challenging and at least as intricate as the parenting practiced in the standard model of social monogamy.

*Cowbirds.* When I was watching the oriole build its nest over a nearly two-week time span, little did I know that beside me a brown-headed cowbird had also been keeping track of it (at a more discreet distance). After the oriole was incubating I climbed to the nest and saw a brown-speckled egg among three of the oriole's. Since there were only three oriole eggs and these birds usually lay four, the cowbird had likely removed one of the oriole eggs. It would be impossible for me to distinguish one baby oriole from another, but each of the three oriole eggs was individually distinct and easy for me to tell apart. Each was inscribed by a different idiosyncratic pattern of straight and squiggly black lines on a background of pale blue. Many of the dark etchings were almost microscopically thin, while others were thick and bold and terminated in big round blobs. What, I wondered, could be the use of such artful design on the eggs at the bottom of a bag of weathered milkweed fibers hanging so high in a tree? Why had the bird evolved to have eggs displaying such apparent extravagance when a raven with its equally colorful egg of green, blue, gray, and black blotches would accept even a hen's egg painted a bright crimson or a potato? Why would the phoebe in the shed nearby accept a cowbird egg among its own pearly white ones, or why would the chipping sparrow in the hedge with its lightly spotted baby blue eggs incubate a cowbird's, huge in comparison and a completely different color? How could cowbirds get away with such blatant nest parasitism when cuckoos' hosts are so acutely discriminating that they have to color-match their eggs almost perfectly to have any chance at all of getting away with their heists?

Cowbirds are New World brood parasites that belong to the family Icteridae, to which orioles, grackles, meadowlarks, bobolinks, and blackbirds also belong. The best-known cowbird species, distributed over the entire North American continent except for the far north, is the brown-headed cowbird. (Two other species are rare in North America and occur farther south.) The brown-headed cowbird has, like the common cuckoo, evolved the evolutionary specialization of dumping its eggs into other birds' nests and never building a nest, never incubating, and never feeding its young. It parasitizes the nests of hundreds of different species. Although they may occasionally lay eggs into artificial nests containing artificial eggs,[40] and I have seen a cowbird examining a year-old vireo nest, cowbirds by no means just wander around dumping eggs into any nests they find at random. Instead, they find nests by watching birds and becoming attuned to their activities.

To have any chance of being incubated to term, parasitic eggs must be laid among fresh eggs. To achieve the critical moment for egg insertion, the cowbird female (and perhaps the male) spends hours watching specific birds, trying to discover them during nest building. After discovering an active nest, they then keep track of it. During this time they are physiologically getting ready to synchronize egg-laying with their prospective hosts. When the host is physiologically ready to lay an egg, then they are too, and when the host has laid the first one or two eggs of its clutch, and before incubation has started, the cowbird rushes in, usually about a half-hour before sunrise. Host species don't lay until shortly after sunrise.[41] Usually the day before, but sometimes the same day or the next day, the cowbird makes a second visit to the host nest to remove a host egg. Defenses of the hosts against the parasitism range from rushing in to sit on the nest and thus cover the eggs, direct attack, throwing the cowbird egg out, burying it in the nest lining, or abandoning the nest.

Some birds, including American robins and catbirds, routinely throw out cowbirds' eggs. Others, like yellow warblers, may abandon their nest or build a new nest lining to cover the parasite (and their own) eggs, starting over. Most birds in North America, however, accept the cowbirds' brown-speckled eggs even when they contrast with their own. The brown-headed cowbird parasitizes well over a hundred host species (141 documented), and although it might seem that most birds are passive recipients of their nest parasitism, there is much evidence that an active arms race exists between them and their hosts.[42] It has, however, played out very differently from that of cuckoos. A cowbird can't just drop eggs here and there into nests. It requires continuous nest monitoring. What cues does it use? It can't just follow any bird, because many migrants—white-throated sparrows, yellow-rumped warblers, Nashville warblers, and ruby-crowned kinglets—are all around when the phoebes lay their eggs. Following these other birds would be futile. Following some resident birds—such as American goldfinch—who are vigorously singing in the spring would also be futile as well, because they don't breed until three months later. Nor would finding any empty nest be any help, because many old nests from the last nesting season are still up. Finding the phoebe nest in the shed and being ready with an egg to lay into it at precisely the right moment indicates that cowbirds keep tabs on other birds before any eggs appear.

In the spring of 2008 I noticed a female cowbird accompanied almost always by a male, and sometimes several, around our house in the last week of April. I made a note of it and kept checking the phoebe nest daily by using a mirror. The first phoebe egg appeared on April 28. Laying one egg per day, she should have laid her fourth egg on May 1. However, that morning as I walked into the shed I found a fresh white phoebe eggshell on the ground by the door. That morning the phoebe didn't sing. Anxiously I checked the nest: it contained three phoebe eggs and one cowbird egg, which I removed. It was plain to see what had happened: the female cowbird had picked the right moment to slip in and insert an egg, and near that same time she had removed one phoebe egg. I do not know if she consumed it to gain nourishment for laying more eggs, removed it to decrease competitors for the food her baby would later need, or removed it to keep the number of eggs in the nest constant so that if the phoebe could count she would not leave her nest when it contained an extra egg. Whatever the reason, the timing was perfect, because I now expected the phoebe to start incubating, and after she did start—either this day or the next—then the nest would be constantly covered and safe from other cowbird sneak attacks.

I thought everything was now on its normal course and nothing more unexpected could happen to my phoebe clutch. However, about two hours later when I was again near the shed, which is centrally located between my little office building, the raven aviary, and our outdoor wood furnace, I heard the phoebe's loud cheeping, and out of the corner of my eye I saw a bird violently chase another out of the woodshed. Had another cowbird tried to parasitize the nest? I eagerly checked the nest again. To my great surprise, the nest now had another phoebe egg missing. A third egg was then added the next day, three phoebe eggs were subsequently incubated, and no second cowbird egg appeared.

The cowbirds, which I had seen near the house every day for weeks, then almost immediately left the premises, so I checked with the neighbors and their phoebes. On one neighbor's property the resident phoebe nest is up on an apparently choice spot on a roof drainpipe, where it has been for many years. I got an extension ladder and looked: five fresh white eggs (fresh white eggs are translucent, incubated eggs are more chalky). Our other neighbor also has a phoebe nest, but hers is in her garage, and it is also deep inside on a ceiling beam where no chipmunk or squirrel can reach it, nor is it visible from outside. She parks her car underneath the nest, and a couple of days after the cowbirds left our nest she noticed an eggshell on her car hood. She thought her phoebes had hatched, but reported that the eggshell was "speckled." That is, a cowbird not a phoebe egg had been expelled from the nest. She then checked the nest and found "four white" (phoebe) eggs and "one speckled" (cowbird) egg. This meant that more than one cowbird had parasitized this nest, and one of them had apparently thrown out the other's egg, because phoebes don't throw eggs out of their nest.

The next spring (2009) "my" phoebes, or another pair, used the same nest. This year I was not particularly interested in the details of what might happen, but when the birds seemed to abandon the nest, I checked the contents. To my surprise it held one cowbird egg, nothing else. Furthermore, the egg had a two-millimeter ragged hole in it, suggesting it had been pecked. I found this very intriguing. It looked as though the cowbird had made the mistake of laying its egg *before* and not after the phoebe laid its egg or eggs. This suggests that maybe phoebes can't distinguish a cowbird egg from their own but "know" that if an egg appears in their nest before they have laid an egg, then it can't be theirs. The bird's inhibitions against damaging an egg in its nest may begin only after it has started laying, because after that any egg in the nest could be its own.

Although parasitization by a brown-headed cowbird does not usually kill the host's young, it may cost the parents one or two fewer young fledged.[43] However, presumably keeping cowbird young in the nest narrows the safety margin when bad weather can kill the whole brood by starvation. Furthermore, baby cowbirds continue to demand the host's attention after fledging and could delay a second brood.

In 2007 the phoebes in our shed had fledged five young and one cowbird on June 13. After that day I saw no more of the young phoebes, but the baby cowbird stayed around the premises and continued to be fed until at least June 30, when it still followed one of the phoebes around the yard and begged loudly. The female phoebe had already started to lay her second clutch on June 20, but did not incubate consistently until June 29. In one study Justin L. Rasmussen and Spencer G. Sealy made 102 observations of forty-five cowbird host species feeding their cowbird fledgling, and in ninety-seven of these observations there were no records of the hosts feeding their own young.[44] They supposed that the host's young were starved, and that therefore the major brood reduction in cowbirds occurs only after the young fledge. However, it is also possible that one member of the host pair leaves with its young into the forest (where cowbirds don't go) to take care of them alone, and the young cowbird, who refuses to go there, stays and begs so that the other pair member must stay to feed it.

There is debate in the bird literature about host defenses against cowbird parasitism, including why the brown-headed cowbird removes an egg of its host's clutch when it parasitizes a nest.[45] The fact that the cowbird risks a second visit to the nest, where it might be attacked, suggests that the behavior is adaptive; that is, it has an evolved benefit because it persists despite a cost. If it is removing competition for food, why does it not remove *more* than one egg? Why does it not just peck them all so that they do not hatch, as another species, the shiny cowbird of southern Florida, does?[46] Why does it not remove all the competition, like cuckoos do? One hypothesis proposed by Rebecca Kilner and colleagues is that it is adaptive for the cowbird to tolerate additional begging mouths in the nest, because that provides a greater stimulus for the parents to bring more food, which the cowbird then intercepts.[47] Do parasitic cuckoos provide much more stimulus simply because of their even bigger and brightly colored mouths, so they don't need to tolerate competitors in the nest?

Cuckoos are often much larger relative to their hosts than cowbirds, and their young require a correspondingly greater share of the food than the host parents bring. Yet, the cuckoo female also removes only one egg when she inserts her own; although the removal of other eggs by the adult female cuckoo would be easy, she still does not do it. Her restraint must have a function. I presume some birds recognize something strange about the new egg, but their own eggs, if left in the nest, are hostages to it, so they don't desert the nest. However, the cuckoo has a counter strategy, activated in the nestling stage *after* the eggs have been accepted and incubated; the cuckoo nestling, who will soon need much more food than any of the host nestlings, eliminates all the others. In an act worthy of an acrobat, the hatchling cuckoo balances any still unhatched host eggs on a flat or hollow shape on its back. It keeps the egg from rolling off to the sides by its enlarged first digit of its wings. Then, while rising up on its legs and bracing itself with its stubby wings for stability, it moves backward until it has elevated its rump to the nest edge and rolls the egg to and over the nest rim. In some species of cuckoo this eviction event does not take place until the host's young have hatched, and in this case the baby cuckoo (as before) has to time it right. It waits until the host mother is absent and until air temperatures are low, when the host nestling is inactive and doesn't put up a struggle.[48] The strategy of shoving the host's eggs or young out of the nest would not work in deep baglike nests and in tree-hole nests. However, some cuckoos (and also the parasitic honey-guides, which parasitize nests that are in tree holes or otherwise so constructed that there is no nest edge that the host's nestlings can reach) have sharp hooks on their bills with which they kill the nestlings. Getting rid of competition is a prime selective pressure of parasitic birds (except maybe cowbirds), and it is especially critical for parasites such as cuckoos, which are much larger and require much more food than their hosts.

Most songbirds' eggs hatch semi-synchronously since incubation begins on the penultimate egg laid or earlier. The majority of the clutch then hatch together, but the last to hatch are at a slight disadvantage and serve as an insurance policy in case food supplies run low or an egg does not hatch. Any extra food requirements, such as to feed a parasite, are a liability to the bird's ability to raise the maximum number of young in every clutch. A cowbird weighs more than a phoebe, and since in some years of our commonly prolonged cold and rainy weather all the babies die of starvation even in unparasitized phoebe nests, any added parasite can push the economic stress beyond the tolerable level—and in the bird world there is never a "stimulus package" to keep them alive.

Cowbirds do not feed their young, but a cowbird's commitment to its egg does not necessarily stop after depositing it in a host's nest. Cowbirds may monitor their chosen nest with their egg in it. In one recent study that aptly titled cowbirds' antics "retaliatory mafia behavior," Jeffrey Hoover and Scott Robinson had been removing cowbird eggs from the nests (in nest boxes) of prothonotary warblers in an effort to increase the warblers' reproduction.[49] However, to their surprise they found *more* nest failures in warbler nests from which they had removed the cowbird eggs than in nests in which they had left cowbird eggs! Suspecting that the cowbirds "punish" hosts who remove their eggs, Hoover and Robinson reduced the nest-box entrance size to exclude cowbirds from the nests they had parasitized but left them still large enough to allow entry of the smaller warblers. These nests then fledged normal broods.

How can cowbirds gain an advantage by their seemingly spiteful behavior of destroying others' broods when their eggs have been evicted? The likely answer is that the host pair whose brood fails immediately (usually within a day or two) starts to make another nesting attempt, so the cowbird gets a second chance to parasitize a nest with fresh eggs. Similarly, a study in Europe had suggested that the common cuckoo also plunders nests that are too advanced for parasitism. The intended hosts then build a fresh nest nearby and the cuckoo parasitizes it in the egg-laying stage.[50] The bronzed cowbirds of Mexico and southern Texas sometimes also destroy all the eggs, but only if they are incubated and the nest can't be parasitized anymore. As with the cuckoo's parasitism, the intended host bird soon starts to build a new nest, and so the cowbird can get a second chance to lay an egg into a fresh clutch. Another hypothesis is that this cowbird destroys the eggs incidentally to assess whether or not the host is suitable for parasitism since incubated eggs indicate that it is too late to leave one of their eggs.[51] If so, then it predicts these are nests that had not been watched.

Finally, the issue of color-matching is not yet settled. It is indeed curious that in the co-evolution of those birds who insert their eggs into the nests of other societies and those potential hosts who detect and discriminate against those parasite eggs, the cuckoos have evolved amazing egg color mimicry, whereas cowbirds have not and yet are at least if not more successful nest parasites. Cuckoos have been in the parasitizing business far longer than cowbirds, and they are therefore assumed to have perfected egg color matching, whereas cowbirds and their hosts are only in the early stages of an arms race. Cuckoos are indeed much more primitive birds than cowbirds (who are a recent offshoot from other icterids), but the dates when either started to become brood parasites are unavailable from the fossil record. How long does it take to evolve color-matching in eggs? And how would anyone know whether it requires 2 or 60 million years? It can't be too long because some species have a great variety of egg color even now. One could also spin natural history scenarios. For example, at least at the present time, egg color matching could be a liability in cowbird brood parasitism, whereas it is a necessity in cuckoo parasitism. The difference relates to details in the *method* of egg insertion.

Cuckoos parasitize a nest by removing a host egg and then staying to insert their own. It's all done in one nest visit. A cowbird, on the other hand, is much smaller than a cuckoo relative to its hosts, and it usually inserts an egg in a hurry, leaves, and comes back much later to remove a host egg. It must therefore be able to recognize its own egg. If it were color-matched, might not the cowbird end up throwing out its own egg by mistake? A cuckoo that is much larger than its hosts is less likely to be intimidated and can, or does, accomplish both acts with one visit. Indeed, it may have to do it in one visit because a cuckoo (with "perfect" egg matching) must, I suspect, remove the host egg before putting in its own, so it won't accidentally eject its own egg from the host's nest.

*Bog-slogging among grackles.* Seeing the grand patterns is a necessary scaffolding for understanding Nature. But ultimately in the specific cases, dozens of grand patterns may converge into a complex picture. In that picture, there is often no one pattern that sticks out. The ecological context, in which everything converges (such as in Charles Darwin's "tangled bank" analogy), delights. The best "tangled bank" that I have is right here in my backyard and I revel in it. I revel in the common grackles and the red-winged blackbirds in my beaver bog, and I puzzle at what may go on there in the nesting season.

Both the common grackles and red-winged blackbirds are "accepter" species of cowbird eggs, and they should be ideal cowbird hosts because of their similarity in size and diet, yet their nests are very seldom parasitized. Over the last fifty years I have looked into at least a hundred redwing nests and dozens of grackle nests in Maine and Vermont and I have never found a nest with a cowbird egg in it. Why not?

Both species have conspicuous open nests that are hard to miss, and in the nesting season and I can find many of them on any day in our bog, where there are always dozens. Proof enough for me that redwings are accepters of cowbird eggs is one (the only one) redwing nest I found in Idaho that contained four cowbird eggs and only two redwing eggs. On the other hand, less than a hundred meters away from the bog where I have examined dozens of redwing nests, cowbirds routinely parasitize my phoebe nest, even when it is built deep inside our shed at a place where it could not possibly be seen from the outside, and scarcely even from the inside. If they had searched, it would have been impossible for cowbirds not to find the grackle and red-winged blackbird nests in my bog.

The eggs of the common grackle and red-winged blackbird are beautifully and variously colored in greens, blues, and browns with lined blotches and black squiggles. The variety of coloring does not seem to make sense in terms of camouflage from predators. The grackles nest in a patch of cattails whose roots have formed a floating matlike surface on the water, and I didn't know how deep the water is below it. There are two cleared beaver channels running through it, and solid bottom there is about five feet down. But I did not want to break through the cattail-root mat and sink out of sight into the muck while I carried my camera. Nevertheless, for photographs of grackle nests I had to risk it.

Grackle nests are bulky and untidy-looking. They resemble oversized robin nests. Like robin nests they have a deep round cup of cementlike consistency that is made with dried mud on a platform of leaves and debris. Each nest is built entirely by the female, who lines it with thin brown fibrous strands. The four to six blue-green eggs, erratically marked in black and brown, look vibrant, almost glowing in their nest cup. As I hunted for their nests in my beaver bog

on June 14, 2008, I found eleven nests woven into last year's collaps-
ing, dead, tan-yellow cattail fronds. The nest bottoms were almost
touching the water and no more than four inches above it. Most were
within four meters of each other, and all were aggregated within a
hundred-meter square area of the bog. I searched in two comparable
areas in the same bog and did not find another nest. In contrast, the
redwings' nests were spread all over the bog. But what surprised me
even more was that, unlike all four phoebe nests that I had just examined (which were synchronized to within two to three days), these
grackles' nests were highly asynchronous in their development. One
nest was just starting to be built, two had the mud cups made but
were still without a lining, another was almost finished, four had
clutches of from four to six eggs, two had just-hatched pink young,
and the last one had five several-day-old young covered in loose black
down.

As before, there were no cowbird eggs in these nests. However,
one clutch of the grackle eggs contained an egg that stood out from
the three others. That clutch had the typical grackle eggs, blue-green
overlaid with diffuse swaths of light brown to give a khaki effect. The
odd egg was a common color morph of the local grackles—clear blue
with crisp sharp black squiggles and violet spots. Suspecting that it
could be one laid by a different female than the nest owner, I briefly
dunked the eggs into the water: the odd egg slowly sank (it was therefore fairly fresh) and the other three stayed up with their tops just
barely breaking the water surface (they were lightly incubated). Had
the odd egg been dropped into the nest by another female, probably a
day after the others had already been laid by the nest owner? Alternately, maybe it was just an odd or last-laid egg, one that didn't happen to be colored like the rest. Could any female risk throwing such
an egg out, even if she could see that it had different coloring?

Although I have not seen a single cowbird egg in any of many
dozens of grackle and redwing nests in this bog over the years, the

parasitized red-wing nest I found in Idaho was alone, in a narrow wash. Here in my Vermont bog, where both grackles and redwings nest in an effective colony, these birds routinely converge like a black swarm whenever a predator (mink, otter, raven, or crow) appears in the marsh and raids their nests. Cowbirds are routinely all around, parasitizing the orioles, phoebes, and other birds. Do grackles and redwings normally chase off any cowbirds that come near? If so, such habits would explain why they are accepters—because they hardly ever have a cowbird egg in their nest! I speculate therefore that cowbirds are indeed normally not allowed into the bog, though this should be tested with at least a dummy mount. The colonial grackles and redwings gang up on all predatory birds that come near. Why not target cowbirds as well?

*Camouflage, and brown eggs versus white.* The original eggs of Ur-birds were probably white. Reptiles, who bury theirs, and those birds who nest in cavities (woodpeckers, hoopoes, motmots, rollers, kingfishers, oil-birds, owls, petrels, swifts, parrots, bee-eaters, barbets, and horn-bills) still are.[52] Several other birds, such as hummingbirds and doves that have open nests, have uncolored or pale eggs as well, so hole nesting as such cannot be the only valid reason for lack of color. However, these other birds lay only two eggs per clutch, and they are incubated (and hence covered) as soon as the first one is laid. Hence they are seldom exposed to the light of day even though they are not in a dark hole. Still other birds, like ducks, geese, and grebes, lay clutches of multiple eggs that are not incubated until the full clutch has been laid, yet they are also unmarked. However, in these birds the eggs are covered by nest material that the female scrapes over the eggs to hide them when she leaves the nest and after each egg is laid. There are, however, also curious examples of eggs that are white and differ from the norm of the group. They include the pure white eggs of Bachman's and Swainson's warblers. All other of the fifty-three North

American wood warblers (Parulidae) have spotted eggs. Both of these warbler species occur in southern swamps and nest close to or on the ground. In Harrison's highly informative book on these birds, the Swainson's warbler has been found so "tame on the nest" that the incubating bird did not flush from the nest, and when his hand was held over the nest, "she straddled my fingers in trying to get back onto it."[53] Might these swamp warblers be subject to heavy predation by snakes, who would hunt eggs not by looking down into nests so that they would see the eggs, but by looking up from the ground? By evolving to sit tight, parents seldom expose the eggs. Similarly, shore birds have fabulously camouflaged eggs, yet only one species has white eggs, the crab plover, and this is also the only shorebird that nests in burrows. Possibly these exceptions prove the rule of specific adaptation.

One of the clearest correlations between birds' natural history and egg coloration is that of the traditional or ancestral cavity nesters who have uncolored eggs and build no nests. The picture becomes murky in cavity-nesting birds that are related to species in groups of traditionally noncavity nesters. They include isolated species of ducks, geese, falcons, corvids, thrushes, pigeons, flycatchers, wrens, finches, starlings, owls, seabirds, and two warblers. In these cases where the group as a whole had nest-building habits and egg coloration, both the nest-building and the egg coloration are maintained despite the species' present nesting in enclosed spaces. I suspect therefore that they still have spotted eggs because they are "recent" hole-nesters who have either experienced low costs to maintain the coloring or little benefit to remove the spotting.[54] Alternately, in cases in which most of the group has colored eggs in open nests, there are a few species now nesting in holes that, unlike the rest of the group, have evolved to have white, unspotted eggs. These include bank and tree swallows. These contrasting patterns emphasize both the major role of color as camouflage to reduce egg predation and also the possibly confusing effect of evolutionary lag that would cloud many specific examples of egg coloring that exist now.

Birds that nest in the open and have several eggs in a clutch that are subject to several days of visual exposure to egg predators invariably have cryptically colored eggs. The concealment afforded by coloration alone is often remarkable. Most people will be familiar with the value of egg camouflage when they encounter the "nest" of a killdeer: a bare scrape on the ground where the eggs are fully exposed yet nearly invisible except when underfoot. Arctic tern eggs on a sandbar with brown cobble and gravel (on the Noatak River in Alaska) were barely visible to me even as I stood over them. Years later while guided by researchers studying a mixed-species tern colony (of common, roseate, and Arctic terns) on the Isles of Shoals in Maine, I again saw Arctic tern nests, but these were among green vegetation. All of these eggs had an olive-green tint, and they again matched their background as much as those did in Alaska on a very different background. That is, at least for Arctic terns, the egg color within one species matched the egg's specific environment, but though variable, color of all three tern species was similar at one location. Both Arctic terns and sandwich terns are known for extreme color variation of the eggs among different individuals, although the reason for it is not understood.[55]

The color codes of eggs arose from the selective pressure of predators and nest parasites in the same way that the selection of predation by birds has produced the beauty and variety of caterpillars, moths, and butterflies, and the competition for pollinators has produced the color and shapes of flowers. However, as much as we may see patterns, the devil is in the details, and ultimately we are interested in the patterns in order to understand the details. The same principles of color coding apply throughout: there is a minority advantage to creating diversity, and that has ultimately created the variety and beauty that delights and the mystery that is all around. Close to my doorstep in the woods nest winter wrens, and they have white eggs with only a few very light flecks of brown, as do sedge wrens. I expect white eggs in both because they are hidden in deep domed nests. But next to the house also nest house wrens, and down in the bog, marsh wrens. Their eggs are also hidden in domed nests, yet both of these have chocolate brown eggs. Why? What are the details that have made these eggs so very different?

I've spent hours sloshing around in a coastal Maine cattail marsh, hearing marsh wrens who were advertising themselves by their thrumming, churring songs. They breed in semicolonial aggregations, and like many wrens the males are polygynous. Their globular nests with a side entrance near the top are composed of dead cattail leaves woven into last year's still-upright dead cattail fronds. These nests, in contrast to winter wren nests, can't be missed, and I find many. However, the side entrance hole in any one nest is not obvious because it is camouflaged by loose vegetation, but when I do find it I poke a finger inside and discover that almost all are without a nest lining. These are "dummy" or display nests built by males to attract females. One or more females may accept these offerings, then finish a nest by putting in the nest lining and laying eggs. I found seven nests in one male's territory, and then found another displaying male with nine such nests. They were close enough together for me to photograph the whole assembly. Of fifteen marsh wren nests that I examined, only one had eggs.

The unusually dark chocolate brown eggs of marsh wrens, which are the darkest of any wren if not of any songbird in North America, challenge expectations.[56] They "should" be white because they are in a dark cavity; there is little chance for light to enter the nest. Furthermore the eggs at the bottom of the deep nest cavity are almost buried in feathers and loose fluff. Since no predator could possibly be attracted to a nest by seeing the eggs, there should be no reason to camouflage them with pigments. Yet there is pigment—and lots of it.

Marsh wren males make many dummy nests and (females?) destroy the eggs and small nestlings of other birds as well as destroying each other's eggs.[57] One hypothesis is that this intolerance is an adaptive behavior because it drives away their competitors for food. But might the two phenomena be connected and reflect competition among the females as well as female mate choice? In Europe, the greater reed warbler, an ecological replacement for the marsh wren, is also polygynous and the females also destroy each other's eggs and small young when they can. In this case, however, a study by Bengt Hansson, Staffan Bensch, and Dennis Hasselquist determined that a male reed warbler has a primary female who receives most of his attention, so that her young had better survival rates than secondary females, who are largely without his help in childrearing unless something happens to the nest of his first wife.[58] These researchers found that whenever a new female settled in a male's territory, there followed a spike in damage to the eggs in other nests in that territory. (The researchers cleverly used dummy eggs that showed beak marks of attempted egg damage.) Apparently, therefore, the second wives compete for the male's help, and they get it by destroying their female competitor's eggs and offspring. Whether the same rationale explains the marsh wren's nest predation is not known, but it would be expected. As follows, I suspect that darkly pigmented eggs should present a challenge for egg dumping, and thus be an adaptive strategy against it.

Marsh wrens breed in relatively uniform expanses of cattails where physical nest sites, cattail plants, grow next to each other and are available in practically unlimited number. A male usually has two and sometimes up to four females. Typically he attracts one female to his dummy nests in one part of his territory, a "courtship site," and after she inspects them and chooses one of his nests, she lines it with brown cattail down. He leaves her then and moves on and builds more dummy nests in another part of his territory to attract another female. The wrens commonly breed close to each other in what appear to be little colonies. Any male may also build up to a dozen dummy nests. Although these nests presumably function proximally as sexual attractants, I suspect, as follows, there is a lot more to this story: perhaps even infanticide and cuckoldry through egg dumping by the females.

An environment with many close neighbors provides opportunity for females to "cuckold" those neighbors by dumping eggs in their nests (or by destroying their eggs), and that means the potential victims must be prepared to resist such incursions. One defense a female might employ would be to choose a male who offers many dummy nests, not to have her pick of the best one, but because of the sheer number. Numerous nests would signify male vigor and confer genetic advantage for her offspring, but there may be a more direct advantage. She chooses only one nest, but the empty nests surrounding hers will make it more difficult for a neighbor to monitor in which one her clutch is located. That is, many dummy nests could allow her to play a shell game in which the chances of being hit are less the more empty nests there are. Furthermore, the potential egg dumper can't just quickly drop an egg into any nest. She can only cuckold a nest where egg laying has just started—one where the resident female is not yet incubating—and in order to do that she would have to closely inspect the contents of numerous nests before finding the right one. She can only afford to drop an egg into a nest if it already contains one to several fresh ones, and the nest owner would not allow her much time to make a detailed search inside her nest. White eggs are not visible from the outside, but they should become visible when the potential cuckolder peeks through the nest entrance, whereas a dark egg would either blend in with the fluffy brown nest lining or else be covered by large feathers. I saw examples of both.

Like marsh wrens, house wrens also have chocolate-colored eggs and also destroy other birds' nests. They nest in cavities and make fake nests in other cavities they find near their own nests. Do these also serve as a deterrent to egg dumping? Sedge wrens have white eggs, but their nest structure is similar to that of marsh wrens, and they provide a curious contrast. The main natural history difference between the two species appears to be that whereas the marsh wrens nest over standing water and hence can reliably be present regardless of major weather fluctuations, sedge wrens wander widely and often nest in pastures and grasslands and in the margins of temporary wetlands. They are rare and are usually present in small numbers, notorious for their erratic movements at seemingly odd times. They have flexible and variable breeding schedules throughout the summer. The end result is that any one nest is unlikely to have a neighbor in synchronous breeding condition who could cuckold or damage. Thus the selective pressure to hide the eggs from other wrens who would try to force their way into a nest is low. The egg's whiteness is thus not significant. It is simply an absence of pigment. However, I suspect white eggs may indeed have functional significance in at least one bird, the tree swallow, in whom it may also relate to nest lining and egg dumping. The riddle includes feathers.

There is a debate in the ornithological literature about why tree swallows line their nests with feathers, and it mainly centers on their function as insulation, possibly as a default because feathers can and do insulate. But that is no argument, any more than the suggestion that we evolved fingers for picking our noses because it is statistically demonstrable that fingers are our primary tool for that purpose. Natural history often provides the hints that can lead to relevant tests.

In most birds, the male builds the basic nest structure as an inducement for a housing start to attract a female, but tree swallows do it in reverse. The female builds a nest of fine dry grasses when she starts laying eggs. Then, near mating time and even during early incubation, the male goes on a feather-collecting binge. He steals feathers from others if he can, and pitched battles between males for a single feather may erupt. The big question is, what are these very hard-to-get feathers good for? If they serve as nest insulation, it still doesn't explain why males primarily collect them, when in most birds it is the female who lines the nest. Part of the published argument for his involvement in feather-collecting to line the nest concerns copulations.

Tree swallows are frequent copulators. Rather than copulating just once, which is all that is normally or potentially necessary to fertilize a clutch of eggs, they copulate about fifty times per clutch of eggs, and they copulate not only with their own mates but also with others' as well; 50 to 87 percent of broods have extra-pair paternity. Copulations are potentially costly. M. P. Lombardo and four co-workers, who manually expressed semen from male swallows at hourly intervals, showed (not unexpectedly!) that increased copulations result in a drastic reduction of the sperm count.[59] So why do it? The argument holds that males copulate numerous times with their mate because females solicit them, inducing them to stay around to feather their nest.[60] Still another hypothesis is that females solicit copulations either so that their mates will not engage in extra-pair copulations or because if they did they would then not care for their young. In any case, apparently these hypotheses don't pan out because as Linda Whittingham and co-workers determined, within-pair copulations are no deterrent for extra-pair copulations.[61] Nor is there evidence that females can successfully coerce males to bring them feathers (or food for their young) by copulating with them. The utility of providing the feathers for insulation sounds feasible and was left by default, but to me it does not explain why feathers are preferred, since many of these feathers are huge, not the small fluffy feathers generally preferred by other birds for insulating their nests. Furthermore, the feathers used by the tree swallows are mostly (but not exclusively) the ones that are hard to get: white ones.

*May 22, 2008.* Our tree swallows have finally started to incubate, but in the last two days they still brought light-colored feathers to their nest. These are eight to nine inches long, and when the birds land at the nest entrance hole, the feathers dangle down even below the tail of the swallow holding it, and the bird is unable to get the feather through the four-centimeter entrance hole. The swallow tries repeatedly, and after failing to get in holding the feather crosswise to the hole, it flies off and tries again a few more times before dropping it to the ground, picking it back up and trying again a few more times. The obsessive efforts to add these long feathers to an already feather-lined nest are curious. Long feathers cannot serve as nest insulation—nests require small downy feathers. I check inside the nest—four white eggs on white contour feathers that curl over the tops of the eggs and almost cover them and the incubating bird.

I counted the feathers in my starling box positioned a few meters from the swallow nest. I found ninety-one brown, nine white, and six black feathers. In contrast, the tree swallow nest in the box next to it, which I also examined after the young fledged, contained seventy-five pure white feathers, twenty-seven mostly gray feathers with white at the bottoms, and eight variously colored feathers—but not one black one. I have examined dozens of swallow nests both in Vermont and in Maine and almost all were lined with predominantly white feathers (although I did not bother to count them). Curiously, one nest with eggs contained no feathers. Perhaps none had been available, or had the male left?

I believe that there is a possible explanation not only for the utility of feathers and their color, but also why it is advantageous for both members of the tree swallow pair to have them in the nest. The first part of my argument is that using white feathers is probably costly, because the birds must search far and wide to find them. It took the swallows three weeks, an extraordinarily long time, to build their flimsy nest. In contrast, cliff swallows can build their nests and a mortared receptacle to hold it in a week, and they nest in large colonies where the mud walls of one nest can be incorporated into the wall of the adjacent nest.[62] On the other hand, a pair of fork-tailed palm swifts, who build a nest of feathers that are glued together with saliva and attached to the ends of palm fronds, require three months to complete their nest, apparently because of the difficulty of securing the feathers; they are procured through pirating by plucking them out of flying birds.[63] The cost to the swallows of using white feathers when any feather, hair, or moss would work perfectly well as insulation argues for a specific function for such feathers.

Tree swallows use only shallow nest holes (whose openings, unlike those of marsh wrens, can't be closed) that are in open areas rather than in dense vegetation. They also face a potentially severe housing shortage, which could be an inducement for them to dump eggs into others' nests, and if they have evolved to dump eggs into each other's nests, then one would also expect there would be an evolved defense. Do the white feathers serve to hide the white eggs?

Tree swallow males do not feed their mates, and only the female incubates, so there is potential for another female to slip in and insert an egg when the nest owner leaves to forage. Generally, however, when the female is about to leave her nest to forage, she perches in the nest-hole entrance as though to plug it or maybe to signal the male to come and replace her at the entrance when she leaves. These behaviors look suspiciously like nest guarding. Perhaps another swallow could slip into the nest to lay an egg, but it would have to be quick to pull it off. A potential egg dumper would have to land on the nest-hole entrance and look down into the nest to assay the presence of fresh eggs. A billowing mass of white feathers would hide the eggs.

With the above natural history considerations in mind, the idea that the swallows might indeed be rejecters of strange eggs, which before had seemed highly unlikely for a hole nester, now suddenly seemed worthwhile enough for me to test, since if they have evolved to become rejecters then it means that they have faced systematic egg dumping in the past. Indeed, the three chickadee eggs that my swallow pair in Maine had apparently discriminated against were highly suggestive. But I needed a better test.

Our swallow pair had just laid their clutch, and so I painted one of their eggs with black and red spots and also added a barn swallow egg and a marsh wren egg. There was no doubt: all of these eggs were rejected just like the chickadee eggs had been—after two days these foreign eggs had been shoved to the side, and the female was brooding only her own eggs.

*Color-coding the mouths of widowfinches.* In the same way that the egg color codes reflect selective pressure of nest parasitization by eggs of the same and other species, so there is also increasing evidence that mouth colors of some nestlings have probably also been shaped by brood parasitization. On the face of it, this may seem like a stretch, because although some birds may be willing to risk rejecting some eggs that may not be their own, they become notoriously forgiving of variation in their young. Indeed, there is conflicting evidence about whether or not the parents care at all about the specific color of their young's maws.[64] A small warbler that accepts and feeds a relatively giant cuckoo with a much larger and brighter red mouth lining, hugely different from its own young's maw, and a cardinal who feeds a fish at the edge of a pool, may make the notion that parent birds note and discriminate among subtle differences in mouth color seem slightly ridiculous. However, these are not necessarily fair examples of the host bird's capacities. A bird that has lost its clutch of young (as in the case of the cuckoo host, as well as probably the cardinal feeding the fish) does not have its own young nearby to compare mouth coloring. Acceptance of young who gape is merely an expression of strong maternal instincts discharged upon the closest prototype available.

Nevertheless, there is more to being a baby bird than just begging vocally, which can be dangerous because it can attract nest predators. Baby birds signal to be fed by stretching their neck and opening their mouth wide to provide a conspicuous target. That target is made all the more obvious by bright coloration—usually pink from the blood. The mouth is framed by glowing yellow or orange "lips" around the edge of the beak from carotenoid pigments. In the majority of songbirds this visual signal tends to be relatively similar among species. Cuckoos tend to have more brightly colored mouths than their hosts, suggesting that their color may be a supernormal stimulus necessary to effectively compete for their foster parent's attention.

A test in discrimination between gapes requires that the alternatives exist side by side, and in brood parasitization in birds a dramatic example is offered in sub-Saharan Africa with estrildid finches (Estrildidae). They are parasitized by indigobirds and whydahs (also called widowfinches) whose babies'mouths are spectacularly colorful and widely divergent between species, yet amazingly convergent between the host estrildid and the parasitic widowfinch in the same nest.

Widowfinch parasitization apparently started as intra-specific parasitization; one finch dumping eggs into the nest of another of its same species since the hosts and parasites are taxonomic relatives. In the evolution of their divergence that started as intra-specific egg dumping, both host and presumptive parasite would have had (and still have) white eggs, so there was little or no variability of egg color for selection to work on to evolve a counter-strategy leading to egg rejection. It should have been relatively easy for these birds to dump eggs into each other's nests, and indeed the widowfinches typically dump not just one but two to four eggs into any one host nest. Selective pressure to eject them should have been strong, but the selective advantage of the parasites to take advantage of the opportunity provided by the presumptive hosts was perhaps even stronger.

The estrildids are a large group of little finches numbering about twenty species. Unlike cowbirds and cuckoos, each species of widowfinch parasitizes only one species of finch. Female widow-finches are plain-colored sparrowlike birds like their hosts, whereas the males in nuptial garb are spectacularly bright indigo (indigobirds), or else they have (in some species) huge long tails that trail out a foot or more behind them in flight. But nothing seems more surprising and incongruous than the stunning color display you see when their nestlings open their mouths and reveal almost luminescent knobs of bright blue, gold, red, black, and white that are arranged in various patterns. Those colors and patterns are different in each species, so that while the young and females of different species are almost indistinguishable in feather plumage, these mouth linings of the babies are species' signatures. Perhaps more surprising, though, is the fact that these mouth linings almost perfectly (to our eyes) match the mouth linings of their host estrildid species. Why?

The first and perhaps most obvious hypothesis of this color-matching is that, in analogy with the egg color matching by cuckoos, these parasites' mouth coloring evolved to mimic their host's because the hosts discriminated against parasite mouths that didn't match. However, this hypothesis is probably incorrect.

The aforementioned cardinal dumping food into the mouth of a fish gaping (possibly for air) poolside may seem bizarre, but not really so when we consider that strong parental instincts can overwhelm many barriers and forgive many flaws in one's offspring. Misidentification is often not the problem of acceptance behavior, instead it is the major component of the solution of successful parenting behavior. And while some birds may be willing to discriminate against some eggs in their nest, once their investment has been great enough to produce young, almost nothing will then make them desert. Sexual reproduction produces variety, and the young change appearance as they grow so that leeway in acceptance must be part of the parenting equation, even though it may come at great cost, discomfort, and even self-sacrifice. If the bonds are not strong, then any excuse might be sufficient to terminate the caregiving. (There is a common misconception that birds will leave their nests if we touch it or the young. Not true. They may, however, easily leave a nest if it is molested while it is being built, while the investment is still small.)

The marvel is not that birds accept strange eggs and hugely mismatched young. The marvel is that some don't accept some eggs. Even mammals are not exempt from what may in special circumstances appear to be misfiring of the parental instinct. In the woods by my camp in Maine a beagle wandered under a brush pile where a bear sow had newborn cubs. She held the dog captive as though it were one of her own cubs, reaching to pull it back in whenever it wanted to leave. The bear had to be shot with a tranquilizer before she could release the unharmed adoptee. Similarly, a sixteen-month-old human toddler of the Lori tribe of nomads in the oak-forested hills of Lorena Province in Iran wandered into a cave where a lactating bear was hibernating, and the sow adopted him.[65] The child was found after three days and "rescued." It smelled of milk; it had been nursed. It was considered a miracle. Undoubtedly this bear would not have given up the child without a fight, so we can presume she was killed for her adoption when it was discovered. I routinely adopt baby animals and bond to them, and vice versa. Our mutual acceptance has nothing to do with misidentification. I usually know pretty well what I'm adopting, and when a raven "adopts" a strange egg in its nest, or the cardinal feeds the fish, it is also not necessarily the consequence of insufficient sensory discrimination. Songbirds of innumerable species feed cuckoo babies they have adopted that are several times their own size. So why then would the mouth linings of some parasitic finches look *precisely* like the complex mouth color pattern of their hosts? The answer is that things are often not what they appear; an experiment shows that although the hosts and parasites have evolved to have mouth colors to look alike, it is not because the parasites mimic their hosts, but vice versa: the hosts mimic their parasites!

In this experiment Justin G. Schuetz from Cornell University used the common waxbill with its almost perfectly mouth-mimicking parasite, the pin-tailed whydah.[66] These hosts readily fed parasite young even after their mouth markings were experimentally altered to differ greatly from their host's. That is, the hosts do not reject non-matching parasite young. They were not even discriminated against in the amounts of food they were provided. Schuetz then speculated that possibly if the parents were handicapped and thus less able to forage they would *then* be forced to discriminate in order to favor their own young. To find out he therefore increased the parent's foraging costs to limit food brought back to the nest (by clipping their flight feathers). Contrary to that prediction, however, the parasites with altered or nonmatching mouth markings then survived even *better* than when the parents were not handicapped! Therefore the parasites provide such a powerful stimulus to entice the hosts to feed them that it is the hosts that adapt to match the parasites' mouth markings, rather than the reverse.[67] The rationale, as follows, stresses the normal adaptive response of bird parenting.

Normally when their brood might starve, bird parents preferentially feed the healthier young because they have a better chance to survive than the possible sick ones who will likely starve, so that investing in them is wasted effort. The weaker the young, the less it begs and the lower the blood flow and paler the mouth lining, thus bright mouth color is a health indicator. That is, greater size, begging vigor, and showiness are signals that parents heed as the young in a nest compete against one another for food. Possibly the parents are attentive to contrast. A parasite in competition for the host's attention to be fed would evolve a bright, contrasting signal that would maximize the attention paid to it. But once a variation in color due to a mutation of a very common brood parasite caught the parent's attention, it could become a model for selection in the hosts. In this case, since the parasites are closely related to the host species, the similar genetic background should have facilitated the same selection and the same response on both species. However, the more precise the matching became between these species, the poorer it was with another species, and the one-on-one species relationship that we now observe could have evolved. But that created two other problems of adaptation, namely host preferences and mating preferences.

In order to maintain the mouth-color matching in the estrildid-vidua host-parasite association for continued co-evolution, it became essential that the vidua parasite reared by any one species of estrildid host also return to parasitize that same species of finch when it becomes an adult. Furthermore, it needs to mate with others who had been reared by the same hosts. Without these behaviors, the genetics of the color coding would dissolve. How, then, would parasites "know" who had parented them so that the female would lay her eggs into the proper species of finch nest? At the same time, she must be attracted to her own kind for mating so that reproductive isolation in this species is maintained, preserving the selected mouth color pattern.[68]

In many birds the young have an innate "template" that allows the males to preferentially learn the songs of their parents and the females to recognize them also and be attracted to them as adults. However, if different widowfinch species themselves had innate species-specific song that acts as a sexual attractant, then they would mate regardless of what host their mate had chosen as foster parents. Thus host matching would be diluted and the parasitization ability which requires specialization would be compromised. The evolved solution, as worked out by Robert Payne and colleagues at the University of Michigan, is that the young widowfinches don't have innate song-learning templates, but instead, while still in the nest, they learn the songs of the males of their estrildid foster "parents."[69] The male widowfinches, when adult, then sing the estrildid song.

In one experiment that clinched this hypothesis, Payne and colleagues reared young African village indigobirds with pairs of the

Bengalese finch, or with pairs of their normal hosts, the firefinch.[70] The males, when adult, in both cases then sang the songs of their male foster parents, and the females from these experiments preferred the males of their foster parents as well. Later, when given a choice, they laid their eggs in the nest of the species that raised them even when the foster species was not the normal host species.

Female indigobirds in the wild hear not only male indigobirds who have learned estrildid songs, they presumably also hear the same songs sung by real estrildid finches. There is more involved with mate choice than sound, or else indigobird females would also be sexually attracted to estrildid males when they sing. They are not, and therefore must require something more—probably the spectacular visual displays and appearance that have evolved in the indigobirds and widowfinch males. These then provide a second and final signal required for species identification by females. Experiments by Malte Anderson, in which he made the normally very long, black, showy tail feathers of male whydahs shorter by clipping a section out, then made the tails of other individuals longer by splicing those sections in with superglue, showed that the shorter-tailed individual males got only half the matings of the long-tails, while the young males, who are drab like the females, got none.[71]

# END OF
# THE NESTING
# SEASON

BY THE END OF AUGUST THE TREES ARE
still in green foliage in Maine and Vermont. The goldenrod,
blue chicory, and white Queen Anne's lace are in full bloom, and
the asters are about to flower. Insect and caterpillar populations
are at their peak. Dragonflies patrol in squadrons and hawk
mosquitoes. Curiously, most of the birds have retreated into the
background. The bogs that had sounded like they hosted a sym-
phony of red-winged blackbirds, marsh wrens, song and swamp

sparrows, yellow warblers, and a bittern, all chiming in every morning and evening, are now silent. Sparrows remain hidden in the tangle of thick growth, but the redwings and the grackles left their summer haunts a month ago. I hear a few cheeps now and then as mixed-species flocks of warblers sweep through the forest. They are molting out of their bright nuptial plumage, punctuating the end of the breeding season. Only the goldfinch males are still in their bright lemon yellow and black nuptial garb, and they are now noisily rearing their young, having put off their nesting season until the thistle seed crop ripened.

It seems too early to quit nesting now since there are still at least two months of warm weather left and "Indian Summer" just around the corner. Now and in the two months to come there is still much more food than there was in early May (and in the case of Canada geese and red-winged blackbirds, early March) when the birds came ready to start nesting almost if not fully from the day they arrived.

In September and early October there is, however, again a stirring of the birds. We see the first long V's of the loudly honking Canada geese pass overhead, all pointing south like compass needles. Incongruously, there is a return of some bird song after a long hiatus. For a day or two, a red-eyed vireo sings from the still-green maples (although a few are starting to show fall colors). He sang only a few notes at a time, never his full song, and always at half-throttle. A winter wren occasionally chimes in with a chirp or two from the woods. Here and there a robin, a phoebe, or a song sparrow sings a few notes like those heard in spring, but then he abruptly stops, as though forgetting his lines. Bluebirds pass through briefly and inspect nest boxes. Pairs of geese drop in to the pond and honk loudly. I suspect some birds are already scoping out where they may return next spring from their winter quarters, if they survive.

*October 24, 2007.* It didn't seem like a particularly promising day when I got up. In the past two days, temperatures soared into the seventies, and people remarked how late the fall was this year. Although most of the red maples have by now long shed their red leaves, the sugar maples' leaves are still up, though bright golden in color. Many of the oaks are still green. But everything changed yesterday after a sharp, cold wind from the northwest loosened the leaves and swept them along the ground. It drove birds south as well. In the daytime, whenever there was a wind from the north, big V's of Canada geese —sometimes hundreds—passed very high overhead, using tailwinds to help push them south. But today the wind stopped, the trees suddenly looked bare, and the sky was dark. In a minute, though, my mood lifted. Could it be that I had just heard an "oogl-la-eee" down in the bog?

Yes, there it was again. Redwings! Their territorial clarion call was repeated again, and again, and again. I had last heard their songs in June when they were nesting there in the cattails. I had last seen them in July, after they had finished raising their young after which they had, as always, left. I ran back to the house, put on a pair of boots, and raced into the bog to see them. There I found the caller, still attired as in spring in his black with crimson shoulder patches. But he was not down in the cattails. Instead, he was facing the bog from the tip top of a pine tree. He kept repeating the call. I noticed a slight hesitation—the last note trailed off slightly, rather than being emphasized as in spring when presumably he and other males first arrived. I then saw five more redwing males, and all of them were perched together in the top of a nearby pine. They also faced into the bog but made only their less inspiring "tick" contact calls and only occasionally voiced an "oogl-la-eee."

An hour later the five left their tree, circled the bog several times, and then settled out of view on the other side. I went back to the house for breakfast, and within minutes one of the birds came up as well, apparently searching at the site where the bird feeder with black sunflower seeds had been in the spring, and where they had come when they first arrived and then visited daily. I suspect that these just-returned redwings were some of the same individuals that had nested here before and that had been coming to my feeder since 2003. I raced out to put the feeder back, but they had already left to return to the bog. Meanwhile, the five males on the pine trees had dropped down into the cattails. Two were perched near each other, performing their full-blown "oogl-la-eee" call/displays and simultaneously bowing and displaying their bright red wing epaulets. Others were spread out over the bog and doing the same thing. The scene looked identical to what I so eagerly await when they first come at the end of March to start their nesting season. As I had written in my notes on March 21, 2003, "Six redwings arrived today, even as the bog is still under deep snow, and the pond ice is almost two feet thick. (I chopped through it all the way to the bottom.) The six spread out over the bog as if each already knew his place. As it got dark, four flew up into the trees, perched for a few minutes, and then flew off into the distance as a group. Two remain and dive down into the cattails." After that the six males were back in the bog every morning. For many days they left in the evening, possibly flying to a nearby communal roost, but returning again in the morning, and their "oogl-la-eee"s reverberated from one end of the bog to the other.

This time these males continued to call back and forth, and after they gathered into a flock at 10:30 AM, they then flew on. They would be heading south, perhaps to Kentucky, Tennessee, or Louisiana. I knew this was the last I would see of them until five months later, at the end of March or early April, when they will return for another nesting season.

Every year I have seen something new in the light of what I had learned before, and so the redwings will undoubtedly once again bring opportunities for new observations that will generate more questions about courting, mating systems, brood parasitism, and parenting. I can't wait.

APPENDIX:

Latin Names of Bird Species

NOTES

ACKNOWLEDGMENTS

INDEX

Appendix:

LATIN NAMES OF

BIRD SPECIES

American coot—*Fulica americana*

American crow—*Corvus brachyrhynchos*

American goldfinch—*Carduelis tristis*

American kestrel—*Falco sparverius*

American redstart—*Setophaga ruticilla*

American robin—*Turdus migratorius*

Arctic tern—*Sterna paradisaea*

Bachman's warbler—*Vermivora bachmanii*

Bald eagle—*Haliaetus leucocephalus*

Bank swallow—*Riparia riparia*

Barbets—*Capitonidae* spp.

Barn owl—*Tyto alba*

Barn swallow—*Hirundo rustica*

Bearded tit—*Panurus biarmicus*

Bearded vulture—*Gypaetus barbatus*

Bee-eaters—*Merops* spp.

Belted kingfisher—*Ceryle alcyon*

Bengalese finch—*Lonchura striata*

Bicknell's thrush—*Catharus bicknellii*

Bishopbird—*Euplectes* spp.

Black-billed cuckoo—*Coccyzus erythrocephalus*

Black-billed magpie—*Pica pica*

Black-capped chickadee—*Parus atricapillus*

Black coucal—*Centropus grillii*

Black-crowned night heron—*Nycticorax nycticorax*

Black-eared cuckoo—*Chrysococcyx osculans*

Black-headed grosbeak—*Pheuticus melanocephalus*

Black-headed gull (Common)—*Larus ridibundus*

Blackpoll warbler—*Dendroica striata*

Black-throated blue warbler—*Dendroica caerulescens*

Black-throated green warbler—*Dendroica virens*

Black woodpecker—*Dryocopus martinus*

Blue-footed booby—*Sula nebouxii*

Blue-headed vireo—*Vireo solitarius*

Blue jay—*Cyanocitta cristata*

Blue tit—*Parus caeruleus*

Blue-winged warbler—*Vermivora pinus*

Bobolink—*Dolichonyx oryzivorus*

Bohemian waxwing—*Bombycilla garrulus*

Boreal chickadee—*Parus hudsonicus*

Broad-winged hawk—*Buteo platypterus*

Bronzed cowbird—*Molothrus aeneus*

Brown-capped weaver—*Phormoplectus insignis*

Brown creeper—*Certhia americana*

Brown-headed cowbird—*Molothrus ater*

Buffalo weaver—*Bubalornis albirustris*

Burrowing owl—*Athene cunicularia*

Cactus wren—*Campylorhynchus brunneicapillus*

California quail—*Callipepla californica*

Canada goose—*Branta canadensis*

Canvasback—*Aythya valissinera*

Cape penduline tit—*Authoscopus miutus*

Cardinal (Northern)—*Cardinalis cardinalis*

Carolina wren—*Thryothorus ludovicianus*

Catbird (Gray)—*Dumetella carolinensis*

Cave swallow—*Petrochelidon fulva*

Cedar waxwing—*Bombycilla cedrorum*

Chestnut-sided warbler—*Dendroica pensylvanica*

Chimney swift—*Chaetura pelagica*

Chinstrap penguin—*Pygoscelis antarcticus*

Chipping sparrow—*Spizella passerina*

Cliff swallow—*Hirundo pyrrhanota*

Coal tit—*Periparus* (formerly *Parus*) *ater*

Collared flycatcher—*Ficedula albicollis*

Common cuckoo—*Cuculus canora*

Common grackle—*Quiscalus quiscala*

Common moorhen—*Gallinula chlorops*

Common murre—*Uria aalge*

Common raven—*Corvus corax*

Common snipe—*Gallinago gallinago*

Common (European) swift—*Apus apus*

Common tern—*Sterna hirundo*

Common waxbill—*Estrilda astrild*

Crab plover—*Dromas ardwola*

Crested tit—*Lophophanes* (formerly *Parus*) *cristatus*

Crested tree swift—*Hemiprocna carunata*

Curlew sandpiper—*Calidris ferruinea*

Desert sparrow—*Passer simplex*

Diederick's cuckoo—*Chrysococcyx caprius*

Downy woodpecker—*Picoides pubescens*

Dunnock—*Prunella modularis*

Dusky shearwater—*Puffinus obscurus*

Eagle owl—*Bubo bubo*

Egyptian plover—*Pluvianus aegypticus*

Emperor penguin—*Aptenodytes fosteri*

European nuthatch—*Sitta europaea*

European starling—*Sturnus vulgaris*

European white stork—*Ciconia ciconia*

Evening grosbeak—*Coccothaustes vespertinus*

Fairy tern—*Sterna nereis*

Fairy wren—*Malarus* spp.

Ferruginous hawk—*Buteo regalis*

Firefinch—*Lagonosticta senegala*

Florida scrub jay—*Aphelocoma coerulescens*

Fox sparrow—*Passerella iliaca*

Giant cowbird—*Scaphidura oryzivora*

Glaucous-winged gull—*Larus glaucescens*

Godwit—*Limosa limosa*

Golden eagle—*Aquila chrysaetos*

Golden plover—*Pluvialis* spp.

Golden-winged warbler—*Vermivora chrysoptera*

Goldfinch (American)—*Carduelis tristis*

Goldfinch (European)—*Carduelis carulis*

Goshawk (Northern)—*Accipiter gentilis*

Great auk—*Plautus impennis*

Great blue heron—*Ardea herodias*

Great crested flycatcher—*Myiarchus crinitus*

Greater flamingo—*Phoenicopterus roseus*

Greater scaup—*Aythya marila*

Great horned owl—*Bubo virginianus*

Great reed warbler—*Acrocephalus arundinaceus*

Great spotted cuckoo—*Clamator glandurius*

Great tit—*Parus major*

Green finch—*Carduelis chloris*

Green sandpiper—*Tringa ochropus*

Green woodpecker—*Picus virdis*

Hairy woodpecker—*Picoides villosus*

Harris's hawk—*Parabuteo unicinctus*

Hermit thrush—*Catharus guttatus*

Herring gull—*Larus argentatus*

Honeycreepers—*Palila* spp.

Honey-guides—*Indicator* spp.

Hoopoe—*Upupa epops*

House finch—*Carpodocus mexicanus*

House martin—*Delichon urbica*

House sparrow—*Passer domesticus*

House wren—*Troglodytes aedon*

Imperial woodpecker—*Campephilus imperialis*

Indigobirds—*Vidua* spp.

Indigo bunting—*Passerina cyanea*

Jackdaw (Eurasian)—*Corvus monedula*

Jacobin cuckoo—*Clamator jacobinus*

Junco (dark-eyed)—*Junco hyemalis*

Jungle fowl—*Gallus gallus*

Kentish plover—*Charadrius alexandrius*

Kestrel (European)—*Falco tinnuculus*

Killdeer—*Charadrius vociferus*

King penguin—*Aptenodytes patagonicus*

Kiskadee (Great)—*Pitangus sulphuratus*

Lapwing—*Vanellus vanellus*

Laysan duck—*Anas laysanensis*

Least flycatcher—*Empidonax minimus*

Lesser flamingo—*Phoeniconaias minor*

Lesser scaup—*Aythya affinis*

Little stint—*Calidris minuta*

Little weaver—*Ploceus luteolus*

Long-tailed manakin—*Chiroxiphia linearis*

Long-tailed tit—*Aegithalos caudatus*

Lucy's warbler—*Vermicora lucine*

Mallard duck—*Anas platyrhynchos*

Malleefowl—*Leipoa ocellata*

Manucode—*Manocodia keraurenii*

Marbled murrelet—*Brachyramphus marmoratus*

Marsh tit—*Parus palustris*

Marsh wren—*Cistothorus palustris*

Meadowlark (Eastern)—*Strurnella magna*

Motmots—*Momotidae*

Mourning dove—*Zenaida macrura*

Mourning warbler—*Opornis philadelphica*

Nashville warbler—*Vermivora ruficapilla*

Northern flicker—*Colaptes auratus*

Northern oriole—*Icterus galbula*

Oilbird—*Steatornis caripensis*

Olive-sided flycatcher—*Contopus borealis*

Osprey—*Pandion haliaetus*

Ostrich—*Struthio camelus*

Oystercatcher—*Haematopus ostralegus*

Palm swift—*Tachornis squamata*

Paradise riflebird—*Ptiloris paradiseus*

Parula warbler (Northern)—*Parula americana*

Penduline tit—*Remiz pendulinus*

Peregrine falcon—*Falco peregrinus*

Phalarope—*Phalaropus* spp.

Phoebe (Eastern)—*Sayornis phoebe*

Pied crow—*Corvus corona*

Pileated woodpecker—*Dryocopus pileatus*

Pine grosbeak—*Pinecola enucleator*

Pine siskin—*Carduelis pinus*

Pin-tailed whydah—*Vidua macroura*

Pinyon jay—*Gymnorhinus cyanocephalus*

Prothonotary warbler—*Protonataria citrea*

Purple finch—*Corpoucus porpureus*

Pygmy falcon—*Polihierax semitorquatus*

Razorbill auk—*Alca torda*

Red-backed fairy wren—*Malurus melanocephalus*

Red-breasted nuthatch—*Sitta canadensis*

Red-cheeked cordonbleu—*Uraeginthus bengalus*

Red crossbill—*Loxia curvirostra*

Red-eyed vireo—*Vireo olivaceus*

Redheaded duck—*Aythya americana*

Redpoll (Common)—*Carduelis flammea*

Redstart—*Phoenicurus phoenicurus*

Red-tailed hawk—*Buteo jamaicensis*

Red-winged blackbird—*Agelaius phoeniceus*

Ring-billed gull—*Larus delawarensis*

Ring dove—*Streptopelia risoria*

Ringed plover (Common)—*Charadrius hiaticula*

Roadrunner (Greater)—*Geococcyx californianus*

Rock dove—*Columba livia*

Rock sandpiper—*Calidris ptilocnemis*

Rollers—Coraciidae

Rook—*Corvus frugilegus*

Roseate tern—*Sterna dougallii*

Rose-breasted grosbeak—*Pheuticus ludovicianus*

Ruby-crowned kinglet—*Regulus calendula*

Ruby-throated hummingbird—*Archilocus colubris*

Ruddy duck—*Oxyura jamaicensis*

Ruff—*Philomachus pugnax*

Ruffed grouse—*Bonasa umbellus*

Rufous-naped wren—*Campylorhynchus rufinucha*

Sage grouse—*Centrocercus urophasianus*

Sanderling—*Calidris alba*

Sandhill crane—*Grus canadensis*

Sandwich tern—*Sterna sandvicensis*

Sapsucker (Yellow-bellied)—*Sphyrapicus varius*

Savannah sparrow—*Passerculus sandwichensis*

Scrub jay—*Aphelcoma coerulescens*

Sea eagle (White-tailed)—*Haliaetus albicilla*

Semipalmated plover—*Charadrius semipalmatus*

Semipalmated sandpiper—*Calidris pusilla*

Sharp-shinned hawk—*Accipiter striatus*

Shiny cowbird—*Molothrus bonariensis*

Snow goose—*Chen caerulescens*

Snowy owl—*Nyctea scandiaca*

Sociable weaver—*Philetaius socius*

Song sparrow—*Melospiza melodia*

Sora—*Perzana carolina*

Speke's weaver—*Ploceus spekei*

Spotted cuckoo—*Clamator glandularis*

Spotted sandpiper—*Actitis macularia*

Spotted towhee—*Pipilo maculatus*

Stonechat—*Saxicola torquata*

Superb starling—*Lamprotornis superbus*

Swainson's hawk—*Buteo swainsoni*

Swainson's thrush—*Catharus ustulatus*

Swainson's warbler—*Limnothlypis swainsonii*

Swamp sparrow—*Melospiza georgiana*

Tailorbird (African)—*Orthotomus metopias*

Tawny owl—*Strix aluco*

Temminck's stint—*Calidris temminckii*

Thick-billed murre—*Uria lomvia*

Thick-billed parrot—*Rhynchopsitta pachyrhyncha*

Thrashers—*Toxostoma* spp.

Tree sparrow (American)—*Spizella arborea*

Tree sparrow (Eurasian)—*Passer montanus*

Tree swallow—*Tachycinata bicolor*

Tufted titmouse—*Parus bicolor*

Turnstone—*Arenaria interpres*

Veery—*Catharus fuscescens*

Village indigobird—*Vidua chalybeate*

Virginia rail—*Rallus limicola*

Wagler's oropendola—*Zarhynchus wagleri*

Wagtail—*Motacilla* spp.

Wandering albatross—*Diomedia exulans*

Western sandpiper—*Calidris mauri*

Whimbrel—*Numenius phaeopus*

White-breasted nuthatch—*Sitta carolinensis*

White-crowned sparrow—*Zonotrichia leucophrys*

White-fronted goose (Greater)—*Anser albifrons*

White stork—*Ciconia ciconia*

White-throated sparrow—*Zonotrichia albicollis*

White-winged crossbill—*Loxia leucoptera*

Widowfinches—*Vidua* spp.

Willow tit—*Parus montanus*

Wilson's warbler—*Wilsonia pusilla*

Winter wren—*Troglodytes troglodytes*

Woodcock (American)—*Scolapax minor*

Wood dove—*Columba oenas*

Wood duck—*Aix sponsa*

Wood peewee (Eastern)—*Contopus virens*

Wood thrush—*Hylocichla mustelina*

Yellow-backed weaver—*Ploceus capitalis*

Yellow-billed cuckoo—*Coccyzus americanus*

Yellow-olive flycatcher—*Tolmomyias sulphurescens*

Yellow-rumped cacique—*Cacicus cela*

Yellow-rumped warbler—*Dendroica coronata*

Yellow thornbill—*Acanthiza nana*

Yellow warbler—*Dendroica petechia*

Zebra finch—*Poephila guttata*

N O T E S

GENERAL REFERENCES

C. Harrison and P. Castell, *Bird Nests, Eggs, and Nestlings of Britain and Europe* (New York: HarperCollins, 1998); C. L. Henderson, *Oology: Ralph's Talking Eggs* (Austin: University of Texas Press, 2007).

INTRODUCTION

1. Having now compared the merits of parenting goslings versus chicks as a way for kids to learn about animals, I would definitely recommend chickens, not geese.

1. LOVE BIRDS

1. H. Fisher, *Why We Love: The Nature and Chemistry of Romantic Love* (New York: Macmillan, 2005); R. J. Sternberg and K. Weis, eds., *The New Psychology of Love* (New Haven: Yale University Press, 2006).

2. J. Alcock, *The Triumph of Sociobiology* (New York: Oxford University Press, 2001), 25–28.

3. B. Heinrich, *The Geese of Beaver Bog* (New York: HarperCollins, 2004).

4. Ibid.

5. P. Zak, Oxytocin is associated with human trustworthiness, *Hormones and Behavior* 48 (2005): 522–527; Zak, The neurobiology of trust, *Scientific American* 298, no. 6 (2008): 88–95.

6. Sternberg and Weis, eds., *New Psychology of Love.*

7. M. Beauregard, J. Courtemanche, V. Pawquette, and E. L. St-Pierre, The neural basis of love, *Psychiatry Research* 172, no. 2 (2009): 93–98; H. E. Fisher, A. Aron, and L. L. Brown, Romantic love: A mammalian brain system for mate choice, *Philosophical Transactions of the Royal Society of London, B Biological Sciences* 361, no. 1476 (2006): 2173–2186; Fisher, Aron, and Brown, Romantic love: An fMRI study of a neural mechanism for mate choice, *Journal of Comparative Neurology* 493 (2005): 58–62.

8. M. F. O'Connor, D. K. Wellisch, A. L. Stanton, N. I. Eisenberger, and M. R. Irwin, Craving love? Enduring grief activates brain's reward center, *Neuroimage* 42, no. 2 (2008): 969–972.

9. Heinrich, *The Geese of Beaver Bog.*

## 2.  MONOGAMY

1. D. F. Westneat, P. W. Sherman, and M. L. Morton, The ecology and evolution of extra-pair copulation in birds, *Current Ornithology* 7 (1990): 330–369; D. F. Westneat and P. W. Sherman, Parentage and the evolution of parental behavior, *Behavioral Ecology* 4 (1993): 66–77.

2. I. Rowley and E. Russell, "Philanderers": A mixed mating strategy in the Splendid Fairy-wren *Malurus splendens, Behavioral Ecology and Sociobiology* 37 (1990): 431–438.

3. C. S. Carter and L. L. Getz, Monogamy and the prairie vole, *Scientific American* 268 (1993): 100–106; W. Wickler and U. Seibt, Monogamy in crustacean and man, *Zeitschrift für Tierpsychologie* 57 (1981): 215–234.

4. J. Alcock, Evolution of mating systems, in *Animal Behavior: An Evolutionary Approach,* 9th ed. (Sunderland, Mass.: Sinauer, 2009).

5. T. L. DeSanto, M. F. Willson, K. M. Bartecchi, and J. Weinstein, Variation in nest sites, nesting success, territory size, and frequency of polygyny in Winter Wrens in northern temperate coniferous forests, *Wilson Bulletin* 115, no. 1 (2003): 29–37.

6.  D. L. Major and C. A. Barber, Extra-pair paternity in first and second broods of eastern Song Sparrows, *Journal of Field Ornithology* 75, no. 2 (2004): 152–156.

7.  D. G. Kleiman, Monogamy in mammals, *Quarterly Review of Biology* 52 (1977): 39–69.

8.  G. W. Barlow, Monogamy in relation to resources, in *The Ecology of Social Behavior*, ed. C. N. Slobodchikoff, 55–79 (New York: Academic Press, 1988).

9.  G. C. Williams, *Adaptation and Natural Selection* (Princeton: Princeton University Press, 1966).

10.  F. R. Gehlbach, *The Screech Owl: Life History, Ecology and Behavior in the Suburbs and Countryside* (College Station: Texas A&M University Press, 2008).

11.  J. H. Crook, The evolution of social organization and visual communication in the weaver birds (Ploceinae), *Behaviour* 10 (1964): S1–S178.

12.  B. Beehler, Adaptive significance of monogamy in the trumpet manucode *Manucodia keraudrenii* (Aves: Paradisaeidae), in *Avian Monogamy*, ed. P. A. Gowaty and D. W. Mock, 83–99, American Ornithological Union, Ornithological Monograph 37 (Lawrence, Kans.: Allen Press, 1985).

13.  J. M. Marzluff and R. P. Balda, Resource and climatic variability: Influences on sociality of two southwestern corvids, in *The Ecology of Social Behavior*, ed. C. N. Slobodchikoff, 255–283 (New York: Academic Press, 1988).

14.  J. L. Dowling and K. E. Omland, Return rates in two temperate breeding orioles (Icterus), *Wilson Journal of Ornithology* 121, no. 1 (2009): 190–193; K. Steenhof and B. E. Peterson, Site fidelity, mate fidelity, and breeding dispersal in American Kestrels, *Wilson Journal of Ornithology* 121, no. 1 (2009): 12–21.

15.  M. H. Reynolds, J. H. Breeden Jr., M. S. Vekasy, and T. M. Ellis, Long-term pair bonds in the Laysan Duck, *Wilson Journal of Ornithology* 121, no. 1 (2009): 187–190.

16.  M. Bochenski and L. Jerzak, Behaviour of the White Stork *Cionia ciconias*: A review, in *The White Stork in Poland: Studies in Biology, Ecology and Conservation*, ed. P. Tryjanowski, T. H. Sparks, and L. Jerzak, 295–324 (Bogucki, Poland: Wydawnictwo Naukome, 2006).

17.  M. Johnson, J. R. Conklin, B. L. Johnson, B. J. McCaffery, S. M. Haig, and J. R. Walters, Behavior and reproductive success of Rock Sandpipers breeding on the Yukon-Kuskokwim River Delta, Alaska, *Wilson Journal of Ornithology* 121 (2): 328–337.

18.  M. Johnson and J. R. Walters, Effects of mate and site fidelity on nest survival of Western Sandpipers *(Calidris mauri)*, *Auk* 125, no. 1 (2008): 76–86.

19.  P. A. Prince, A. G. Wood, T. Barton, and J. P. Croxall, Satellite tracking of Wandering Albatross *(Diomedea exulans)* in the South Atlantic, *Antarctic Science* 4, no. 1 (1992): 31–36.

20.   W. L. N. Tickell, *Albatrosses* (New Haven: Yale University Press, 2000).

21.   S. Akesson and H. Weimerkirsch, Albatross long-distance navigation: Comparing adults and juveniles, *Journal of Navigation* 58 (2005): 365–373; F. Bonadonna, C. Bajzak, S. Benhamou, K. Igloi, P. Joventin, H. P. Lipp, and G. Dell'Omo, Orientation in the wandering albatross: Interfering with magnetic perception does not affect orientation performance, *Proceedings of the Royal Society of London B* 272 (2005): 489–495.

22.   P. Jouventin, A. Charmantier, M.-P. Dubois, P. Jarne, and J. Bried, Extra-pair paternity in the strongly monogamous Wandering Albatross *Diomedea exulans* has no apparent benefits for females, *Ibis* 149, no. 1 (2007): 67–78; H. I. Fisher, Experiments on homing in Laysan Albatrosses, *Diomedea immutabilis, Condor* 73, no. 4 (1971): 389–400.

23.   G. S. Fowler, Stages of age-related reproductive success in birds: Simultaneous effects of age, pairbond duration and reproductive experience, *American Zoologist* 35 (1995): 318–328; J. M. Black, Introduction: Pair bonds and partnerships, in *Partnerships in Birds: The Study of Monogamy*, ed. J. M. Black, 3–20 (New York: Oxford University Press, 1996).

24.   J. S. Kellam, Pair bond maintenance in Pileated Woodpeckers at roost sites during autumn, *Wilson Bulletin* 115, no. 2 (2003): 186–192.

25.   L. A. M. Tranquilla, P. P-W. Yen, R. W. Bradley, B. A. Vanderkist, D. B. Lank, N. R. Parker, M. Drever, L. W. Lougheed, G. W. Kaiser, and T. D. Williams, Do two murrelets make a pair? Breeding status and behavior of Marbled Murrelet pairs captured at sea, *Wilson Bulletin* 115, no. 4 (2003): 374–381.

26.   J. C. Bednarz, Cooperative hunting in Harris' Hawks, *Science* 239 (1988): 1525–1527; G. G. Frye and R. P. Gerhardt, Apparent cooperative hunting in Loggerhead Shrikes, *Wilson Bulletin* 113 (2001): 462–464; R. Bowman, Apparent cooperative hunting in Florida Scrub-jays, *Wilson Bulletin* 115, no. 2 (2003): 197–199.

27.   T. J. Cade and G. L. MacLean, Transport of water by adult sandgrouse to their young, *Condor* 69 (1967): 323–343.

28.   G. L. MacLean, Field studies on the sandgrouse of the Kalahari Desert, *Living Bird* 7 (1968): 209–235; MacLean, Adaptations of sandgrouse for life in arid lands, *Proceedings of the 16th International Congress of Ornithology* (1974): 502–516.

29.   T. R. Howell, *Breeding Biology of the Egyptian Plover,* Pluvianus aegypticus, University of California Publications in Zoology 113 (Berkeley: University of California Press, 1979).

30.   I. Szentirma and T. Szekely, Diurnal variation in nest material use by the Kentish Plover *Charadrius alexandrius, Ibis* 146 (2004): 535–537.

31.   D. N. Jones, R. W. R. J. Dekker, and C. S. Roselaar, eds., *The Megapodes* (Oxford: Oxford University Press, 1995).

32.  R. Liverside, The biology of the Jacobin Cuckoo, *Clamator jacobinus*, *Ostrich* 8 (1971): S117–S137.

33.  Stanley Cramp, *Handbook of the Birds of Europe, the Middle East and North Africa* (New York: Oxford University Press, 1985) 4: 391–400.

34.  Robert B. Payne, *The Cuckoos* (New York: Oxford University Press, 2005), 121.

## 3.  POLYGYNY AND POLYANDRY

1.  D. B. McDonald and W. K. Potts, Cooperative display and relatedness among males of a lek-mating bird, *Science* 266 (1994): 1030–1032.

2.  J. Jukema and T. Piersma, Permanent female mimics in a lekking shorebird, *Biology Letters* 2 (2006): 161–164.

3.  B. Hansson, S. Bensch, and D. Hasselquist, Infanticide in great reed warblers: secondary females destroy eggs of primary females, *Animal Behaviour* 54, no. 2 (1997): 297–304.

4.  F. A. Pitelka, R. T. Holmes, and S. F. McLean, Ecology and evolution of social organization in Arctic sandpipers, *American Zoologist* 14 (1974): 185–204.

5.  L. W. Oring, Avian polyandry, *Current Ornithology* 3 (1985): 309–351; L. W. Oring and M. L. Knudson, Monogamy and polyandry in the Spotted Sandpiper, *Living Bird* 11 (1973): 59–73.

6.  L. W. Oring, R. C. Fleischer, J. M. Reed, and K. E. Marsden, Cuckoldry through stored sperm in the sequentially polyandrous Spotted Sandpiper, *Nature* 359 (1992): 631–633.

7.  J. D. Reynolds, Mating systems and nesting biology of the Red-necked Phalarope *Phalaropus lobatus*, *Ibis* 129 (1987): 225–242.

8.  W. Goymann, A. Wittenzellner, and J. C. Wingfield, Competing females and caring males, polyandry and sex-role reversal in African Black Coucals, *Centropus grillii*, *Ethology* 110 (2004): 807–823.

9.  W. Goymann, B. Kempenaers, and J. Wingfield, Breeding biology, sexually dimorphic development and nestling testosterone concentrations of the classically polyandrous African Black Coucal, *Centropus grillii*, *Journal Ornithologie* 146 (2005): 314–324.

10.  Goymann, Wittenzellner, and Wingfield, Competing females and caring males.

11.  S. T. Emlen and L. W. Oring, Ecology, sexual selection and the evolution of mating systems, *Science* 197 (1977): 215–223; N. B. Davies and A. Lundberg, Food distribution and a variable mating system in the Dunnock *Prunella modularis*, *Journal of Ani-*

*mal Ecology* 53 (1984): 895–912; N. B. Davies, I. R. Hartley, B. J. Hatchwell, and N. E. Langmere, Female control of copulations to maximize male help: A comparison of polygynandrous Alpine Accentors *Prunella collaris* and Dunnocks, *P. modularis, Animal Behaviour* 51 (1996): 27–47; M. Nakamura, Multiple mating and cooperative breeding in polygynous Alpine Accentors, *Behaviour* 55 (1998): 259–289; J. D. Ligon, *The Evolution of Avian Breeding Systems* (Oxford: Oxford University Press, 1999).

12.  J. E. Goetz, K. P. McFarlane, and C. C. Rimmer, Multiple paternity and multiple male feeders in Bicknell's Thrush *(Catharus bicknelli), Auk* 120, no. 4 (2003): 1044–1053.

13.  Ibid.

14.  A. M. Strong, C. C. Rimmer, and K. P. McFarlane, Effect of prey biomass on reproductive success and mating strategy of Bicknell's Thrush *(Catharus bicknelli),* a polygynandrous songbird, *Auk* 121, no. 2 (2004): 446–451.

15.  D. R. Rubenstein, The secret lives of starlings, *Natural History* (July/August 2009): 28–33.

## 4. PENGUINS AND US

1.  J. Alcock, *Animal Behavior: An Evolutionary Approach,* 9th ed. (Sunderland, Mass.: Sinauer, 2009).

2.  M. Zuk and N. W. Bailey, 2008. Birds gone wild: Same-sex parenting in albatross, *Trends in Evolution and Ecology* 12 (2008): 658–680.

3.  E. M. Thomas, *The Harmless People,* rev. ed. (New York: Vintage Books, 1989).

## 5. FINE-TUNING NESTING TIME

1.  E. Gwinner, Circannual clocks in avian reproduction and migration, *Ibis* 138 (1996): 47–63; E. Gwinner and J. Dittani, Endogenous reproductive rhythms in a tropical bird, *Science* 249 (1990): 906–908.

2.  P. D. Hunt and B. C. Aliason, Blackpoll Warbler, *Dendroica striata,* in *The Birds of North America,* no. 431, ed. A. Poole and F. Gill (Philadelphia: The Birds of North America, Inc., 1999).

3.  Ibid.

4.  R. Greenberg and P. P. Marra, eds., *Birds of the World: The Ecology of Migration* (Baltimore: John Hopkins University Press, 2005).

5.  B. Helm, T. Piersma, and H. van der Jeugd, Sociable schedules: Interplay between avian seasonal and social behavior, *Animal Behaviour* 72 (2006): 245–262.

6.  H.-C. Schaefer, G. W. Eshiamwata, F. B. Munyekenye, and K. Bohning-Gaese, Life-history of two African Sylvia warblers: Low annual fecundity and long-fledging care, *Ibis* 146 (2004): 427–437.

7.  A. F. Skutch, *Parent Birds and Their Young* (Austin: University of Texas Press, 1976); A. F. Skutch, Clutch size, nestling success, and predation on nests of Neotropical birds, *Ornithological Monographs* 36 (1985): 575–594.

8.  E. Garcia-Vigon, P. J. Cordero, and J. P. Veiga, Exogenous testosterone in female Spotless Starlings reduces their rate of extrapair offspring, *Animal Behaviour* 76, no. 2 (2008): 345–353.

9.  A. D. S. Lehrman, P. N. Brody, and R. P. Wortis, The presence of the mate and nesting material as stimulus for development of incubation behavior and gonadotropin secretion in the Ring Dove *(Streptopelia risoria)*, *Endocrinology* 68 (1961): 507–516; L. Fusani, M. Gahr, and J. B. Hutchinson, Aromatase inhibition reduces specifically one display of the Ring Dove courtship behavior, *General and Comparative Endocrinology* 122 (2001): 23–30.

10.  C. M. Vleck, N. A. Mays, J. W. Dawson, and A. R. Goldsmith, Hormonal correlates of paternal correlates and helping behavior in cooperatively breeding Harris's Hawks, *Parabuteo unicinctus*, *Auk* 108 (1991): 638–648.

## 6. STRUTTING THEIR STUFF

1.  John Burroughs, *Birds and Poets* (Boston: Houghton Mifflin, 1877), p. 87.

2.  M. Nice, *The Watcher at the Nest,* illus. Roger Tory Peterson (New York: Dover Publications, 1967).

3.  J. R. Krebs, The significance of song repertoires: The Beau Geste hypothesis, *Animal Behaviour* 25 (1977): 475–478; J. R. Krebs, Song and territoriality in the great tit, in *Evolutionary Ecology,* ed. B. Stonehouse and C. M. Perrins, 47–62 (London: Macmillan, 1977); P. K. McGregor, J. R. Krebs, and C. M. Perrins, Song repertoire and lifetime reproductive success in the Great Tit *(Parus major)*, *American Naturalist* 118 (1981): 149–159. Quotation from W. A. Searcy and S. Nowicki, Bird song and the problem of honest communication, *American Scientist* 96 (2008): 114–121.

4.  W. A. Searcy and P. Marler, A test for responsiveness to song structure and programming in female sparrows, *Science* 213 (1981): 926–928; A. L. O'Loghlen and M. D. Beecher, Sexual preferences for mate song types in female Song Sparrows, *Animal Behaviour* 53 (1997): 835–841.

5. J. M. Reid, P. Arcese, A. L. E. V. Cassidy, S. M. Hiebert, J. N. M. Smith, P. K. Stoddard, A. B. Marr, and L. F. Keller, Song Sparrow repertoire size predicts initial mating success in male Song Sparrows *(Melospiza melodia), Animal Behaviour* 68 (2004): 1055–1063.

6. J. M. Reid, P. Arcese, A. L. E. V. Cassidy, S. M. Heibert, J. N. M. Smith, P. K. I. Stoddard, A. B. Marr, and L. K. Keller, Fitness correlates of song repertoire size in free-living Song Sparrows *(Melospiza melodia), American Naturalist* 165 (2005): 299–310.

7. Searcy and Nowicki, Bird song and the problem of honest communication; I. F. MacDonald, B. Kempster, L. Zanette, and S. A. Macdonald-Shackleton, Early nutritional stress impairs development of song-control brain regions in both male and female juvenile Song Sparrows *(Melospiza melodia), Proceedings of the Royal Society of London B* 273 (2006): 2559–2564.

8. J. W. Bradbury, Lek mating behavior in the hammerheaded bat, *Zeitschrift für Tierpsychologie* 45 (1977): 225–255.

9. W. A. Searcy, R. C. Anderson, and S. Nowicki, 2006. Bird song as a signal of aggressive intent, *Behavioral Ecology and Sociobiology* 60 (2006): 234–241.

10. M. D. Beecher, Function and mechanisms of song learning in song sparrows, *Advances in the Study of Behavior* 38 (2008): 167–226.

11. A. M. D. Beecher, P. K. Stoddard, S. E. Campbell, and C. L. Horning, Repertoire matching between neighboring song sparrows, *Animal Behaviour* 51 (1996): 917–923; P. K. Stoddard, M. D. Beecher, S. E. Campbell, and C. L. Horning, Song-type matching in the Song Sparrow, *Canadian Journal of Zoology* 70 (1992): 1440–1444.

12. G. L. Patricelli, S. W. Coleman, and G. Borgia, Male Satin Bowerbirds, *Ptilonorhynchos violaceus,* adjust their display intensity in response to female startling: An experiment with robotic females. *Animal Behaviour* 71 (2006): 49–59.

13. K. Stastny, *Vogel: Handbuch und Fuhrer der Vogel Europas* (Augsburg, Germany: Natur Verlag in Westbild Verlag, 1992).

14. G. Borgia, Why do bowerbirds build bowers? *Scientific American* 83 (1995): 542–548.

15. E. Leichty and J. W. Grier, Importance of facial pattern to sexual selection in Golden-winged Warbler *(Vermivora chrysoptera), Auk* 123 (2006): 962–966.

16. D. Klem Jr., Preventing bird-window collisions, *Wilson Journal of Ornithology* 121, no. 2 (2009): 314–321.

17. W. R. Lindsay, M. S. Webster, C. W. Varian, and H. Schwabl, Plumage colour acquisition and behaviour are associated with androgens in a phenotypically plastic tropical bird, *Animal Behaviour* 77 (2009): 1525–1532.

18. G. E. Hill, Plumage coloration is a sexually selected indicator of male quality, *Nature* 350 (1991): 337–339.

19. G. E. Hill, *A Red Bird in a Brown Bag: The Function and Evolution of Ornamental Plumage Coloration in the House Finch* (New York: Oxford University Press, 2002); O. Voelker, Die Abhangigkeit der lipochrombildung bei Voegeln von pflanzlichen Carotinoiden, *Journal Ornithologie* 82 (1934): 439.

20. T. Flinn, J. Hudon, and D. Derbyshire, Tricks exotic shrubs do: When Baltimore Orioles stop being orange, *Birding* (Sept./Oct. 2007): 62–68.

21. W. D. Hamilton and M. Zuk, Heritable true fitness and bright birds: A role for parasites? *Science* 218 (1982): 384–386.

22. M. Zuk, K. Johnson, and J. D. Ligon, Parasites and male ornaments in free-ranging and captive jungle fowl, *Behaviour* 114 (1990): 232–248; A. Møller and P. Anders, Female choice selects for male sexual tail ornaments in the monogamous swallow, *Nature* 332 (1988): 640–642; H. G. Smith and R. Montgomerie, Sexual selection and the tail ornaments of North American Barn Swallows, *Behavioral Ecology and Sociobiology* 28 (1991): 195–202; J. Merila, B. C. Sheldon, and K. Lindstrom, Plumage brightness in relation to haematozoan infections in the greenfinch *Carduelis chloris:* Bright males are good bet, *Ecoscience* 6 (1999): 12–18; G. E. Hill, K. L. Farmer, and M. L. Beck, The effect of mycoplasmosis on carotenoid plumage coloration in male House Finches, *Journal of Experimental Biology* 207 (2004): 2095–2099.

23. K. T. McGraw, P. M. Nolan, and O. L. Crino, Carotenoid accumulation strategies for becoming a colorful House Finch: Analysis of plasma and liver pigments in wild moulting birds, *Functional Ecology* 20, no. 4 (2006): 678–688.

24. A. Zahavi, Mate selection—A selection for a handicap, *Journal of Theoretical Biology* 111 (1975): 205–213; A. Zahavi and A. Zahavi, *The Handicap Principle: A Missing Piece of Darwin's Puzzle* (Oxford: Oxford University Press, 1977).

25. R. D. Estes, Social organization of the African Bovidae, in *Proceedings of an International Symposium on the Behavior of Ungulates and Its Relation to Management,* IUCN Special Publication No. 24, 166–205 (Morges, Switzerland: IUCN, 1974).

26. M. Andersson, Female choice selects for extreme tail length in a widowbird, *Nature* 299 (1982): 818–820; A. Moller and P. Anders, Female choice selects for male sexual tail ornaments in the monogamous swallow, *Nature* 332 (1988): 640–642; H. G. Smith and R. Montgomerie, Sexual selection and the tail ornaments of North American Barn Swallows, *Behavioral Ecology and Sociobiology* 28 (1991): 195–202.

27. N. T. Burley and R. Symanski, A taste for the beautiful: Latent aesthetic mate preference for white crests in two species of Australian grass finches, *American Naturalist* 152 (1998): 792–802.

28. N. T. Burley, P. G. Parker, and K. Lund, Sexual selection and extra-pair fertiliza-

tions in a socially monogamous passerine, the Zebra Finch *(Taeniopygia guttata)*, *Behavioral Ecology* 7 (1996): 218–226.

29. A. Qvarnstrom, V. Blomgren, C. Wiley, and N. Svedin, Female Collared Flycatchers learn to prefer males with an artificial novel ornament, *Behavioral Ecology* 15, no. 4 (2004): 543–548.

30. W. J. Hamilton and R. Poulin, The Hamilton and Zuk hypothesis revisited: A meta-analytical approach, *Behaviour* 134 (1997): 299–320.

31. L. Carroll, *Through the Looking Glass* (London: Macmillan, 1872).

32. M. Milinski, The major histocompatibility complex, sexual selection, and mate choice, *Annual Review of Ecology and Systematics* 37 (2006): 159–186.

33. Hamilton and Zuk, Heritable true fitness and bright birds.

34. R. Ekblom, S. A. Saether, M. Grahm, P. Fiske, J. A. Kalas, et al., Major histocompatibility complex variation and mate choice in a lekking bird, the Great Snipe *(Gallinago media)*. *Molecular Biology* 13 (2004): 3821–3828.

35. M. Petrie, Improved growth and survival of offspring of peacocks with more elaborate tails, *Nature* 371 (1994): 598–599; C. Bonneaud, J. Mazuc, O. Chastel, H. Westerdahl, and G. Sorci, Terminal investment induced by immune challenge and fitness traits associated with major histocompatibility complex in the House Sparrow, *Evolution* 58 (2004): 2823–2830; C. Bonneaud, O. Chastel, P. Federici, H. Westerdahl, and G. Sorci, Complex MHC-based mate choice in wild passerines, *Proceedings of the Royal Society B* 273 (2006): 1111–1116.

36. G. Miller, *The Mating Mind: How Sexual Choice Shaped the Evolution of Human Nature* (New York: Doubleday, 2000).

## 7. NEST SITE AND SAFETY

1. D. I. King and R. M. DeGraaf, Predators at bird nests in a northern hardwood forest in New Hampshire, *Journal of Field Ornithology* 77, no. 3 (2006): 239–243.

2. B. T. Thomas, Family Staetornithidae (Oilbird), in *Handbook of the Birds of the World*, vol. 15, ed. J. del Hoyo, A. Elliott, J. Sargatal, et al. (Barcelona: Lynx Edicions, 1999).

3. T. G. Pearson, ed., *Birds of America* (Garden City, NY: Garden City Books, 1917), sec. 3, p. 165.

4. D. S. Wilcove, Border traffic, *Living Bird* (Winter 2008): 14–16.

5. H. H. Harrison, *Wood Warbler's World* (New York: Simon and Schuster, 1984), 215.

6.  B. A. Hahn and E. D. Silverman, Social cues facilitate habitat selection: American Redstarts establish breeding territories in response to song, *Biology Letters* 2, no. 3 (2006): 337–340.

7.  M. Betts, A. S. Hadley, N. Rodenhouse, and J. J. Nocera, Social information trumps vegetation structure in breeding-site selection by a migrant songbird, *Proceedings of the Royal Society B* 275 (2008): 2257–2263.

8.  H. Kruuk, Predators and anti-predator behaviour of the Black-headed Gull (*Larus ridibundus* L.), *Behaviour* 11 (1964): S1–S129; J. Picman, S. Pribil, and A. Isabelle, Antipredation value of colonial nesting in Yellow-headed Blackbirds, *Auk* 119, no. 2 (2002): 461–472; E. F. Perry and D. E. Andersen, Advantages of clustered Least Flycatchers in North-central Minnesota, *Condor* 105, no. 4 (2003): 756–770.

9.  M. Hansell, *Bird Nests and Construction Behaviour* (New York: Cambridge University Press, 2000).

10.  K. L. Kosciuch and K. A. Arnold, Novel nesting behavior in Cave Swallows, *Wilson Bulletin* 115, no. 3 (2003): 347–348.

11.  R. B. Payne, *The Cuckoos* (New York: Oxford University Press, 2005), 163.

12.  C. Spottiswoode, E. Herrmann, O. A. E. Rasa, and C. W. Sapsford, Co-operative breeding in the Pygmy Falcon *(Polihierax semitorquatus), Ostrich* 75, no. 4 (2004): 322–324.

13.  J.-P. Tremblay, G. Gauthier, D. Lepage, and A. Desrochers, Factors affecting nesting success in the Greater Snow Geese: Effects of habitat and association with Snowy Owls, *Wilson Bulletin* 109 (1997): 449–461.

14.  S. Weidensaul, *Raptors: The Birds of Prey* (New York: Lyons and Burford, 1996), 129.

15.  Ibid, 130.

16.  V. Dethier, *The Ecology of a Summer House* (Amherst: University of Massachusetts Press, 1984).

17.  Harrison, *Wood Warbler's World.*

18.  *Audubon* (March/April 2008).

19.  J. K. Leiser and M. Itzkowitz, The benefits of dear enemy recognition in three-contenders Convict Cichlid *(Cichlasoma nigrofasciatum)* contests, *Behaviour* 136, no. 8 (1999): 983–1003.

20.  P. H. Gosse, quoted in *Costa Rican Natural History,* ed. D. H. Janzen (Chicago: University of Chicago Press, 1983).

21.  D. H. Janzen, *Pseudomyrmex ferruginea* (Acacia ant), in Janzen, ed., *Costa Rican Natural History.*

22. N. G. Smith, The advantage of being parasitized, *Nature* 219 (1968): 690–694; N. G. Smith, *Zarhynchus wagleri* (Chestnut-headed Oropendola), in Janzen, ed., *Costa Rican Natural History.*

23. F. J. Joyce, Nesting success of Rufous-naped Wrens *(Campylorhynchus rufinucha)* is greater near wasp nests, *Behavioral Ecology and Sociobiology* 32 (1993): 71–77.

24. P. Beier and A. I. Tungbani, Nesting with the wasp *Ropalidia cincta* increases nest success of Red-cheeked Cordonbleu *(Uraeginthus bengalus)* in Ghana, *Auk* 123 (2006): 1022–1037.

25. E. S. Morton, J. Howlett, N. C. Kopysh, and I. Chiver, Song ranging in incubating Blue-headed Vireos, *Journal of Field Ornithology* 77, no. 3 (2006): 291–301.

## 8. NEST MATERIALS AND CONSTRUCTION

1. J. R. Horner, *Digging Dinosaurs* (New York: Harper and Row, 1990).

2. D. Lack, *Swifts in a Tower* (London: Methuen, 1956), 66.

3. M. Hansell, *Bird Nests and Construction Behaviour* (New York: Cambridge University Press, 2000).

4. N. E. Collias and E. C. Collias, Evolution of nest building in the Weaverbirds, *Evolution* 19, no. 2 (1965): 267.

5. L. Clark and J. R. Mason, Use of nest material as insecticidal and anti-pathogenic agents by the European Starling, *Oecologia* 67 (1985): 169–176; L. Clark, Countering parasites and pathogens, *Parasitology Today* 6 (1991): 358–360; P. T. Fauth, D. G. Krementz, and J. E. Hines, Ectoparasitism and the role of green nesting material in the European Starling, *Oecologia* 88 (1991): 22–29.

6. R. Pinxten and M. Eens, Polygyny in the European Starling: Effects on female reproductive success, *Animal Behaviour* 40 (1990): 1035–1047; L. Brouwer and J. Komdeur, Green nesting material has a function in mate attraction in the European Starling, *Animal Behaviour* 67 (2004): 539–548.

7. L. Lafuma, M. M. Lambrechts, and M. Raymond, Aromatic plants in bird nests as a protection against blood-sucking flying insects? *Behavioral Processes* 56 (2001): 113–120; H. Gwinner, The function of green plants in nests of European Starlings *Sturnus vulgaris*, *Behaviour* 134 (1997): 337–351.

8. H. Gwinner and S. Berger, European starlings: Nestling condition, parasites and green nest material during the breeding season, *Journal of Ornithology* 146 (2005): 365–

371; H. Gwinner, M. Oltrogge, L. Trost, and U. Nienaber, Green plants in starling nests: Effects on nestlings, *Animal Behaviour* 59 (2000): 301–309.

9. K. Stastny, *Vogel: Handbuch und Fuhrer der Vögel Europas* (Augsburg, Germany: Natur Verlag in Westbild Verlag, 1992).

10. C. R. Brown and M. B. Brown, *Coloniality in the cliff swallow: The effect of group size on social behavior* (Chicago: University of Chicago Press, 1996).

11. M. F. Spreyer and E. H. Bucher, Monk Parakeets *(Myiopsitta monachus)*, *Birds of North America* 322 (1998): 1–23.

12. G. A. Bartholomew, F. N. White, and T. R. Howell, The significance of the nest of the Social Weaver *Philetairus socius:* Summer observations, *Ibis* 118 (1976): 402–411.

13. F. N. White, G. A. Bartholomew, and T. R. Howell, The thermal significance of the nests of the Sociable Weaver *Philetairus socius:* Winter observations, *Ibis* 117 (1975): 171–179.

## 9. THE EGG

1. J. L. Newbrey, W. L. Reed, S. P. Foster, and G. L. Zander, Laying-sequence variation in yolk carotenoid concentrations in eggs of Yellow-headed Blackbirds *(Xanthocephalus xanthocephalus)*, *Auk* 125, no. 1 (2008): 124–130.

2. N. Saino, V. Bertacche, R. P. Ferrari, R. Martinelli, A. P. Moller, and R. Stradi, Carotenoid concentration in Barn Swallow eggs is influenced by laying order, maternal infection and paternal ornamentation, *Proceedings of the Royal Society of London B* 269 (2002): 1729–1733.

3. K. Zimmermann and J. M. Hippner, Egg size, eggshell porosity, and incubation period in the marine bird family Alcidae, *Auk* 124, no. 1 (2007): 307–315.

4. P. D. Boersma, G. A. Rebstock, and D. L. Stokes, Why penguins eggshells are thick, *Auk* 121, no. 1 (2004): 148–155.

## 10. PARENTING IN PAIRS

1. D. W. Mock, Infanticide, siblicide and avian mortality, in *Infanticide: Comparative and Evolutionary Perspectives,* ed. G. Hausfater and S. B. Hrdy, 3–30 (Chicago: Aldine, 1984); Mock and G. A. Parker, *The Evolution of Sibling Rivalry* (Oxford: Oxford University Press, 1997); C. S. Forbes and Mock, A tale of two strategies: Life-history aspects of family strife, *Condor* 102, no. 1 (2000): 23–34.

2. D. W. Mock and L. S. Forbes, The evolution of parental optimism, *Trends in Ecology and Evolution* 10 (1995): 130–134.

3. E. Gwinner, S. Koenig, and C. S. Haley, 1995. Genetic and environmental factors influencing clutch size in equatorial and temperate zone stonechats (*Saxicola torquata axillaris* and *S. t. rubicola*): An experimental study, *Auk* 112, no. 3 (1995): 748–755.

4. B. Helm, E. Gwinner, and L. Trost, Flexible seasonal time and migratory behavior: Results from stonechat breeding program, *Annals of the New York Academy of Science* 1046 (2005): 216–227.

5. R. E. Hegner and J. C. Wingfield, Effects of brood-size manipulations on parental investment, breeding success, and reproductive physiology by House Sparrows, *Auk* 104 (1987): 470–480; T. Slagsvold, Clutch size variation of birds in relation to nest predation: On the cost of reproduction, *Journal of Animal Ecology* 53 (1984): 945–953.

6. C. Frith and B. M. Beehler, *The Birds of Paradise* (New York: Oxford University Press, 1998).

7. C. Dijkstra, A. Bult, S. Bijlsma, S. Daan, T. Meijer, and M. Zijlstra, 1990. Brood size manipulation in the kestrel *(Falco tinnuculus):* Effects on offspring and parental survival, *Journal of Animal Ecology* 59 (1990): 269–285; Meijer, Daan, and M. Hall, Family planning in the kestrel *(Falco tinnuculus):* The proximate control of covariance of laying date and clutch size, *Behaviour* 114 (1990): 117–136.

8. E. Roskaft, The effect of enlarged brood size on the future reproductive potential of the rook, *Journal of Animal Ecology* 54 (1985): 255–260; L. Gustafsson and T. Pärt, Acceleration of senescence in the Collared Flycatcher *Fidecula albicollis* by reproductive cost, *Nature* 347 (1992): 279–281; J. Moreno, R. J. Cowie, J. J. Sanz, and R. S. R. Williams, Differential response by males and females to brood manipulation in the Pied Flycatcher *Fidecula hypoleuca:* Energy expenditure and nesting diet, *Journal of Animal Ecology* 64 (1995): 721–732.

9. N. Tinbergen, The shell menace, *Natural History* 72 (August 1963): 28–35.

10. D. E. Parmelee, Snowy Owl *(Nyctea scandiaca),* in *Birds of North America,* no. 10, ed. A. Poole and F. Gill (Philadelphia: Academy of Natural Sciences and Washington, D.C.: American Ornithologist's Union, 1992).

11. A. Margalida, J. Bertran, J. Boudet, and R. Heredia, Hatching asynchrony, sibling aggression and cannibalism in the Bearded Vulture *Gypaetus baratus, Ibis* 146 (2004): 386–393.

12. M. J. West and A. P. King, Mozart's starling, *American Scientist* 78 (1990): 106–114.

13. C. A. Ristau, Aspects of the cognitive ethology of an injury-feigning bird, the Piping Plover, in *Cognitive Ethology: The Minds of Other Animals,* ed. C. A. Ristau (Hillsdale, N.J.: Lawrence Erlbaum Associates, 1991).

14.  B. Heinrich, *The Geese of Beaver Bog*.

15.  B. Heinrich, *A Year in the Maine Woods* (Reading, Mass.: Addison-Wesley, 1994), 33.

16.  B. T. Thomas, Family Staetornithidae (Oilbird), in *Handbook of the Birds of the World*, vol. 15, ed. J. del Hoyo, A. Elliott, J. Sargatal, et al. (Barcelona: Lynx Edicions, 1999).

17.  C. Haupt, *Mauersegler in Menschenland*, a pamphlet produced by the "Deutsche Gesellschaft fuer Mauersegler e. V." for promoting help in care for the swifts, *Apus apus* (2001).

18.  M. K. Tarburton and E. Kaiser, Do fledgling and pre-breeding Common Swifts *Apus apus* take part in aerial roosting? *Ibis* 143 (2001): 255–263.

19.  J. Holmgren, Roosting in tree foliage by Common Swifts *Apus apus*, *Ibis* 146 (2004): 404–416.

20.  Tarburton and Kaiser, Do fledgling and pre-breeding Common Swifts *Apus apus* take part in aerial roosting?

21.  W. A. Montevecchi, Eggshell removal and nest sanitation in Ring Doves. *Wilson Bulletin* 86, no. 2 (1974): 136–143.

22.  G. Dell'Omo, E. Alleva, and C. Carere, Parental recycling faeces in the Common Swift, *Animal Behaviour* 56, no. 3 (1998): 631–637.

23.  West and King, Mozart's starling.

24.  E. Glueck, Why do parent birds swallow the feces of their nestlings? *Cellular and Molecular Life Sciences* 44, no. 6 (1988):

25.  K. McGowan, A test of whether economy or nutrition determines fecal sac ingestion in nesting corvids, *Condor* 97, no. 1 (1995): 50–56.

## 11.  CUCKOLDS, CUCKOOS, COWBIRDS, AND COLOR CODES

1.  M. M. Kasumovic, L. M. Ratcliffe, and P. T. Boag, Habitat fragmentation and paternity in Least Flycatchers, *Wilson Journal of Ornithology* 121, no. 2 (2009): 306–313.

2.  D. F. Westneat and I. R. K. Stewart, Extra-pair paternity in birds: Causes, correlates, and conflicts, *Annual Review of Ecology, Evolution, and Systematics* 345 (2003): 365–396.

3.  M. Schwartz, D. J. Boness, C. M. Schaeff, P. Majluf, E. A. Perry, and R. C. Fleis-

cher, Female-solicited extrapair matings in Humboldt Penguins fail to produce extra-pair fertilizations, *Behavioral Ecology* 10, no. 3 (1999): 242–250.

4.  S. Hamao and D. S. Saito, Extrapair fertilization in the Black-browed Reed Warbler *(Acrocephalus bistrigiceps)*: Effects on mating status and nesting cycle of cuckolded males, *Auk* 122 (2005): 1086–1096.

5.  M. D. Jennions and M. Petrie, Why do females mate multiply? A review of the genetic benefits, *Biological Reviews* 75 (2000): 21–64; K. Forester, K. Delhey, A. Johnson, J. T. Lifjeld, and B. Kempenaers, Females increase offspring heterozygosity and fitness through extra-pair matings, *Nature* 425 (2003): 714–717; D. F. Westneat, P. W. Sherman, and M. L. Morton, The ecology and evolution of extra-pair copulations in birds, in *Current Ornithology*, vol. 7, ed. D. M. Power, 331–369 (New York: Plenum Press, 1990); S. B. Smith, M. S. Webster, and R. T. Holmes, The heterozygosity theory of extra-pair mate choice in birds: A test and a cautionary note, *Journal of Avian Biology* 36, no. 2 (2005): 146–154; J. L. Brown, The theory of mate choice based on heterozygosity, *Behavioral Ecology* 8 (1997): 60–65; J. H. Zeh and D. W. Zeh, The evolution of polyandry I: Intragenomic conflict and genetic incompatibility, *Proceedings of the Royal Society of London B* 263 (1996): 1711–1717.

6.  M. S. Webster, K. A. Tarvin, E. M. Tuttle, and S. Pruett-Jones, Reproductive promiscuity in the Splendid Fairy-wren: Effects of group size and auxiliary reproduction, *Behavioral Ecology* 15, no. 6 (2004): 907–915.

7.  G. Michl, J. Torok, S. C. Griffith, and B. C. Sheldon, Experimental analysis of sperm competition mechanisms in a wild bird population, *Proceedings of the National Academy of Sciences USA* 99 (2002): 5466–5470.

8.  S. M. Smith, Extra-pair copulation in the Black-capped Chickadee: The role of the female, *Behaviour* 107 (1988): 15–23; Smith, Reflections on a lifetime of ornithological research, *Wilson Bulletin* 115, no. 1 (2003): 73–90.

9.  E. M. Gray, Female Red-winged Blackbirds accrue material benefits from copulating with extra-pair males, *Animal Behaviour* 53, no. 3 (1997): 625–639; M. A. Colwell and L. W. Oring, Extra-pair mating in the Spotted Sandpiper: A female mate acquisition tactic, *Animal Behaviour* 38 (1989): 675–684.

10.  B. Kempenaers, R. B. Lanctot, and R. J. Robertson, Certainty of paternity and paternal investment in Eastern Bluebirds and Tree Swallows, *Animal Behaviour* 55 (1998): 845–860.

11.  M. Osorio-Beristain and H. Drummond, Male boobies expel eggs when paternity is in doubt, *Behavioral Ecology* 12, no. 1 (2001): 16–21.

12.  M. L. Evans, B. J. M. Stutchbury, and B. E. Woolfended, Off-territory forays and mating system of the Wood Thrush *(Hylocichla mustelina)*, *Auk* 125, no. 1 (2008): 67–75.

13. N. B. Davies, Polyandry, cloacal pecking and sperm competition in the Dunnock, *Nature* 302 (1983): 334–336; Davies, Cooperation and conflict among Dunnocks, *Prunella modularis, Animal Behaviour* 33 (1985): 628–648.

14. O. E. Bray, J. J. Kenelli, and J. L. Guarino, Fertility of eggs produced on territories of vasectomized Red-winged Blackbirds, *Wilson Bulletin* 87 (1975): 187–195.

15. P. W. Sherman and M. L. Morton, Extra-pair fertilization in mountain White-crowned Sparrows, *Behavioral Ecology and Sociobiology* 22 (1988): 413–420.

16. P. H. Wrege and S. T. Emlen, Biochemical determination of parental uncertainty in White-fronted Bee-eaters, *Behavioral Ecology and Sociobiology* 20 (1987): 153–160; D. F. Westneat, Extra-pair fertilizations in a predominantly monogamous bird: Genetic evidence, *Animal Behaviour* 35 (1987): 877–886.

17. S. T. Emlen and P. H. Wrege, Forced copulation and intra-specific parasitism: Two costs of social living in the White-fronted Bee-eater, *Ethology* 71 (1986): 2–29.

18. M. A. Fournier and J. E. Hines, Nest sharing by a Lesser Scoup and a Greater Scoup, *Wilson Bulletin* 108 (1996): 380–381.

19. P. W. Govoni, K. S. Summerville, and M. D. Eaton, Nest sharing between an American Robin and a Northern Cardinal, *Wilson Journal of Ornithology* 121, no. 2 (2009): 424–426.

20. B. Semel, P. W. Sherman, and S. M. Byers, Effects of brood parasitism and nest-box placement on Wood Duck breeding ecology, *Condor* 90 (1988): 920–930; Semel and Sherman, Intraspecific parasitism and nest-site competition in Wood Ducks, *Animal Behaviour* 61 (2001): 787–803.

21. A. R. Bower and D. J. Ingold, Intraspecific brood parasitism in the Northern Flicker. *Wilson Bulletin* 116, no. 1 (2004): 94–97.

22. B. E. Lyon, Egg recognition and counting reduce cost of avian conspecific brood parasitism, *Nature* 422 (2003): 495–499; Lyon, Mechanisms of egg recognition in defenses against conspecific brood parasitism: American Coots *(Fulica americana), Behavioral Ecology and Sociobiology* 61 (2007): 455–463.

23. A. P. Møller, Intraspecific nest parasitism and antiparasite behaviour in swallows, *Hirundo rustica, Animal Behaviour* 35 (1987): 247–254.

24. Ibid.

25. Ibid.; H. B. Weaver and C. R. Brown, Brood parasitism and egg transfer in Cave Swallows *(Petrochelidon fulva)* and Cliff Swallows *(P. pyrrhonota)* in South Texas, *Auk* 121 (2005): 1122–1129.

26. C. P. Ortega and A. Cruz, Mechanism of egg acceptance by marsh-dwelling blackbirds, *Condor* 90, no. 2 (1988): 349–358; B. D. Peer and S. G. Sealy, Fate of grackle

(*Quiscalus* spp.) defenses in the absence of brood parasitism: Implications for long-term parasite-host coevolution, *Auk* 121 (2004): 1172–1186.

27. E. C. Collias, Egg measurements and coloration throughout life in the Village weaverbird, *Ploceus cucullatus*, in *Proceedings of the Fifth Pan-African Ornithological Congress* (1984): 461–475.

28. S. Freeman, Egg variability and conspecific nest parasitism in the *Ploceus* weaverbirds, *Ostrich* 59, no. 2 (1988): 49–369; J. K. Victoria, Clutch characteristics and egg discriminative ability of the African Village Weaverbird *Ploceus cucullatus*, *Ibis* 114 (1972): 367–376.

29. P. J. Baicich and C. J. C. Harrison, *Nests, Eggs, and Nestlings of North American Birds* (Princeton: Princeton University Press, 2005); Rosamund Purcell, Linnae S. Hall, and René Corado, *Egg & Nest* (Cambridge, Mass.: Harvard University Press, 2008).

30. B. Tschanz, Zur Brutbiologie der Trottellumme *(Uria aalge aalge Pont.) Behaviour* 14 (1959): 1–100; Tschanz, Trottellummen, *Zeitschrift für Tierpsychologie* 4 (1968): S1–S99.

31. S. G. Sealy, Laying times and a case of conspecific nest parasitism in the Black-billed Cuckoo, *Journal of Field Ecology* 74, no. 3 (2003): 257–260.

32. R. B. Payne, *The Cuckoos* (New York: Oxford University Press, 2005).

33. N. B. Davies, *Cuckoos, Cowbirds and Other Cheats* (London: T. and A. D. Poyser, 2000); N. B. Davies and M. L. Brooke, An experimental study of co-evolution between the cuckoo, *Cuculus canorus*, and its hosts. II. Host egg markings, chick discrimination and general discussion, *Journal of Animal Ecology* 58 (1989): 225–236; I. Wyllie, *The Cuckoo* (London: Batsford, 1981).

34. A. Antonov, B. G. Stokke, A. Moksnes, and E. Roskaft, Egg rejection in Marsh Warblers *(Acrocephalus palustris)* heavily parasitized by Common Cuckoos *(Cuculus canorus)*, *Auk* 123, no. 2 (2006): 419–430.

35. D. C. Lahti and A. R. Lahti, How precise is egg discrimination in weaverbirds? *Animal Behaviour* 63, no. 6 (2002): 1135–1142.

36. D. Attenborough, *The Life of Birds* (London: BBC Books, 1998), 250.

37. R. B. Payne, The ecology of brood parasitism in birds, *Annual Review of Ecology and Systematics* 8 (1977): 1–28.

38. Ibid.

39. M. Soler and J. G. Martinez, Is egg-damaging behavior by Great Spotted Cuckoos an accident or an adaptation? *Behavioral Ecology* 11, no. 5 (2000): 495–501.

40. B. R. Coppedge, Repeated Brown-headed Cowbird parasitism of an artificial nest, *Wilson Journal of Ornithology* 121, no. 1 (2009): 177–180.

41. F. G. McMaster, D. L. H. Neudorf, S. G. Sealey, and T. E. Pitcher, A comparative analysis of laying times in passerine birds, *Journal of Field Ornithology* 75, no. 2 (2004): 113–208.

42. H. Friedmann, L. F. Kiff, and S. I. Rothstein, A further contribution to knowledge of the host relations of the parasitic cowbirds, Smithsonian Contributions to Zoology no. 235, 75 (Washington, D.C.: Smithsonian Institution Press, 1977); Friedmann, Further information on the host relations of the parasitic cowbird, *Auk* 88 (1971): 239–255; Rothstein, Mechanisms of avian egg-recognition: Additional evidence for learned components, *Animal Behaviour* 26 (1978): 671–677.

43. S. K. Davis, Nesting ecology of mixed-grass prairie songbirds in southern Saskatchewan, *Wilson Bulletin* 115, no. 2 (2003): 119–130.

44. J. L. Rasmussen and S. G. Sealy, Hosts feeding only Brown-headed Cowbird fledglings: Where are the host fledglings? *Journal of Field Ecology* 77, no. 3 (2006): 269–279.

45. B. D. Peer, S. I. Rothstein, M. J. Kuehn, and R. C. Fleischer, Host defenses against cowbird (*Molothrus* spp.) parasitism: Implications for cowbird management, *Ornithological Monographs* 57 (2005): 84–97; Peer, Egg destruction and egg removal by avian brood parasites: Adaptiveness and consequences, *Auk* 123, no. 1 (2006): 16–22.

46. A. A. Astie and J. C. Reboreda, Costs of egg punctures and parasitism by Shiny Cowbirds *(Molothrus bonariensis)* at Creamy-bellied Thrush *(Turdus amaurochalinus)* nests, *Auk* 123, no. 1 (2006): 23–32.

47. R. M. Kilner, J. R. Madden, and M. E. Hauber, Brood parasitic cowbird nestlings use host young to procure resources, *Science* 305, no. 5685 (2004): 877–879.

48. G. P. Dement'ev and N. A. Gladkov, eds., *Birds of the Soviet Union*, vol.1 (Jerusalem: Israel Program for Scientific Translation, 1966); R. B. Payne and L. L. Payne, Nestling eviction and vocal begging behaviors in Australian glossy cuckoos *Chrysococcyx basalis* and *C. lucidus*, in *Parasitic Birds and Their Hosts*, ed. S. I. Rothstein and S. K. Robinson, 152–169 (New York: Oxford University Press, 1998).

49. J. P. Hoover and S. K. Robinson, Retaliatory mafia behavior by a parasitic cowbird favors host acceptance of parasitic eggs, *Proceedings of the National Academy of Sciences USA* 104, no. 11 (2007): 4479–4483.

50. N. B. Davies and M. L. Brooke, Cuckoos versus Reed Warblers: Adaptation and counteradaptation, *Animal Behaviour* 36 (1988): 262–284.

51. V. Massoni and J. C. Reboreda, Egg puncture allows shiny cowbirds to assess host egg development and suitability for parasitism, *Proceedings of the Royal Society of London B* 266, no. 1431 (1999): 1871–1874.

52. R. M. Kilner, The evolution of egg colour and patterns in birds, *Biological Review* 81 (2006): 383–406.

53.  H. H. Harrison, *Wood Warbler's World* (New York: Simon and Schuster, 1984); W. M. Jackson, Estimating conspecific nest parasitism in the Northern Masked Weaver based on within-female variability in egg appearance, *Auk* 109, no. 3 (1992): 435–443.

54.  A. G. Gosler, J. P. Higham, and S. J. Reynolds, Why are bird's eggs speckled? *Ecology Letters* 8 (2005): 1105–1113.

55.  Baicich and Harrison, *Nests, Eggs, and Nestlings of North American Birds*.

56.  Ibid.

57.  J. Picman, S. Pribil, and A. Isabelle, Antipredation value of colonial nesting in Yellow-headed Blackbirds, *Auk* 119, no. 2 (2002): 461–472; Picman, Intraspecific nest destruction of eggs by the Long-billed Marsh Wren *(Telmatodytes palustris), Canadian Journal of Zoology* 55 (1977): 1997–2003; Picman, Experimental study on the role of intra- and interspecific competition of nest-destroying behavior in Marsh Wrens, *Canadian Journal of Zoology* 62 (1984): 2353–2356; Picman and J. C. Belles-Isles, Intraspecific egg destruction in Marsh Wrens: A study of mechanisms of preventing filial ovicide, *Animal Behaviour* 35 (1987): 236–246.

58.  B. Hansson, S. Bensch, and D. Hasselquist, Infanticide in Great Reed Warblers: Secondary females destroy eggs of primary females, *Animal Behaviour* 54, no. 2 (1997): 297–304.

59.  M. P. Lombardo, K. L. Green, P. A. Thorpe, M. R. Czarnowski, and H. W. Power, Repeated sampling affects Tree Swallow semen characteristics, *Journal of Field Ornithology* 75, no. 4 (2004): 394–403.

60.  M. P. Lombardo, Within-pair copulations: Are female Tree Swallows feathering their own nest? *Auk* 112, no. 4 (1995): 1077–1079.

61.  L. A. Whittingham, P. O. Dunn, and R. J. Robertson, Do female Tree Swallows guard their mates by copulating frequently? *Animal Behaviour* 47 (1994): 994–997; Whittingham, Dunn, and Robertson, Do males exchange feathers for copulations in Tree Swallows? *Auk* 112, no. 4 (1995): 1079–1080; Whittingham and Dunn, Survival of extrapair and within-pair young in Tree Swallows, *Behavioral Ecology* 12 (2001): 496–500.

62.  Møller, Intraspecific nest parasitism and antiparasite behaviour in swallows; H. B. Weaver and C. R. Brown, Brood parasitism and egg transfer in Cave Swallows *(Petrochelidon fulva)* and Cliff Swallows *(P. pyrrhonota)* in South Texas, *Auk* 121 (2005): 1122–1129.

63.  B. M. Whitney, "Kleptoptily": How the Fork-tailed Palm-swift feathers its nest, *Auk* 124, no. 2 (2007): 712–715.

64.  R. M. Kilner, Mouth color is a reliable signal of need in begging canary nest-

lings, *Proceedings of the Royal Society of London B* 264 (1997): 963–968; D. G. Noble, N. B. Davies, I. R. Hartley, and S. B. McRae, The red gape of the nestling cuckoo *(Cuculus canoris)* is not a supernormal stimulus for three common hosts, *Behaviour* 136 (1999): 759–779.

65. B. Kingsolver, Small wonder, *Orion* (Summer 2002): 1–3.

66. J. G. Schuetz, Low survival of parasite chicks may result from their imperfect adaptation to hosts rather than expression of defenses against parasitism, *Evolution* 59, no. 9 (2005): 2017–2024; Schuetz, Reduced growth but not survival of chicks with altered gape patterns: Implications for the evolution of nestling similarity in a parasite finch, *Animal Behaviour* 70 (2005): 839–848.

67. M. E. Hauber and R. M. Kilner, Coevolution, communication, and host chick mimicry in parasitic finches: Who mimics whom? *Behavioral Ecology and Sociobiology* 61 (2007): 497–503.

68. J. Nocolai, Der Brutparasitismus der Viduinae als ethologishes Problem: Praegungsphaenomene als Factoren der Rassen- und Art-bildung, *Zeitschrift für Tierpsychologie* 21 (1964): 129–204.

69. R. B. Payne, L. L. Payne, and J. L. Woods, Song learning in brood parasitic indigobirds, *Vidua chalybeata:* Song mimicry of the host species, *Animal Behaviour* 55 (1988): 1537–1553.

70. R. B. Payne, L. L. Payne, J. L. Woods, and M. D. Sorenson, Imprinting and the origin of parasitic-host associations in brood-parasitic indigobirds, *Vidua chalybeata, Animal Behaviour* 59 (2000): 69–81.

71. M. Anderson, Female choice selects for extreme tail length in a widowbird, *Nature* 299 (1982): 818–820.

ACKNOWLEDGMENTS

I thank John Alcock and an anonymous reviewer for reading a draft of the manuscript and generously offering advice and numerous suggestions and pertinent literature that I found very helpful and often necessary. Others, perhaps too numerous to all recall, contributed through discussion, providing useful anecdotes and in some cases offering material support. They include Mathew Aeberhard, Ninon Ballerstaedt, Virginia Barlow, Helmut v. Benda, Peter Berthold, Herbert Biebach, William Burt, Anthony Diamond, Toby Fulweiler, Wolfgang Goymann, Renate Grahmann-Opalka, Thomas Gruenkorn, Helga Gwinner, Christy and George Happ, Lynn Havsall, Barbara Helm, Joanne Ignewski, Alain Jacot, Frank J. Joyce, John Marzluff,

Meg McVey, Manfred Milinski, Ruth O'Leary, Robert B. Payne, Noah Perlut, Albert Reingewirtz, Carlos D. Santos, Hubert Schwabl, Nuria Selva, Gregory Septon, Paul Sherman, Avishai Shmida, Adam Wajrak, John Wright. My special thanks go to the editorial staff of Harvard University Press: Michael Fisher, Ann Downer-Hazell, and Kate Brick, who saw this book from the beginnings and through production. They inspired, encouraged, and guided the process seamlessly. Lisa Roberts designed the book. And last but not least I thank Kerry Hardy for a cuckoo roadkill used as a model for one of my watercolor sketches.